The Walkable City

This book explores everyday walking in contemporary urban life. It brings together important theoretical and empirical insights to understand how the 'walkability' of urban spaces can be imagined, planned for, and experienced.

The book focuses on the everyday experiences of the urban walker, the bodily experiences of walking, and different walking research methods. It goes beyond the conventional focus on walkable places by delving into the ways in which urban space is consumed and produced through different ways of walking. Drawing on fieldwork in the UK and international secondary sources, the book examines how walking is socially and materially co-produced, focusing on pedestrian practices, infrastructures, and the social nature of walking. Chapters in the book offer key explorations of the cultural and social inclusions and exclusions of navigating the city on foot. The book considers transport planning and policy promoting pedestrian movement, pedestrian infrastructures, the politics of walking, and social interactions of urban pedestrians. The book offers vital analyses of how different but overlapping dimensions of walking and their relationship with urban space are often overlooked, and the importance of centring the lived experiences of walking in understandings of pedestrian practices.

This book provides a timely contribution to the field of mobilities due to a growing interest in urban walking. It will be of interest to students and scholars of urban studies, human geography, sociology, and public health.

Jennie Middleton is Associate Professor in Human Geography in the Transport Studies Unit and the School of Geography and the Environment at the University of Oxford. She is a human geographer with interests in everyday mobilities, care in the city, and innovative methodologies for urban research.

Routledge Studies in Urbanism and the City

For more information about this series, please visit https://www.routledge.com/Routledge-Studies-in-Urbanism-and-the-City/book-series/RSUC

The Walkable City
Dimensions of Walking and Overlapping Walks of Life

Jennie Middleton

Routledge
Taylor & Francis Group

LONDON AND NEW YORK

First published 2022
by Routledge
2 Park Square, Milton Park, Abingdon, Oxon OX14 4RN

and by Routledge
605 Third Avenue, New York, NY 10158

Routledge is an imprint of the Taylor & Francis Group, an informa business

British Library Cataloguing-in-Publication Data
A catalogue record for this book is available from the British Library

Library of Congress Cataloging-in-Publication Data
A catalog record has been requested for this book

ISBN: 978-1-138-69771-3 (hbk)
ISBN: 978-1-032-05536-7 (pbk)
ISBN: 978-1-315-51921-0 (ebk)

Typeset in Times New Roman
by KnowledgeWorks Global Ltd.

For David and Sue

Contents

List of Figures

Preface

As an academic with an interest in walking, as both a topic and method, I have been asked many times about the types of walking I undertake, where, and how often, with the implicit assumption being that it is both an activity I do and enjoy. People are often surprised, or in some cases shocked, to learn that I don't particularly enjoy walking, nor does it form a significant part of my everyday life where I can help it! Many of my experiences of walking have failed to meet the romanticised visions of walking as a convenient, healthy, and sustainable mode of transport promoted in policy arenas or as the emancipatory urban practice it is positioned as across much of the academic literature concerned with cities. Writing this book has led me to critically reflect upon how my past pedestrian experiences have shaped, and emerge out of, my current everyday mobilities, and how cycling has become the primary means by which I travel on a daily basis.

During my doctoral and postdoctoral studies in London, I walked not from a desire to experience the city on foot but as a means to save money. My everyday journey to the tube station could be done much quicker on the bus but walking enabled me to save the bus fare. Laden with a heavy laptop and books, I would have much preferred to be sat reading on the bus as opposed to pounding the streets with an aching back! This feeling of being encumbered when walking continued when I became a mother. As I trudged through the streets of Oxford in the depths of winter with my first born and in the (sometimes) blistering heat of summer three years later with my second, I willed them to sleep through the continuous motion of pushing the pram as I walked. I did not feel emancipated as I walked through the city but encumbered, exhausted, and full of angst and anticipation of the everyday encounters which might hinder their sleep: the drilling of roadworks, a shouting child, obstacles that hindered continuous movement such as red lights, and any number of other combinations of endless possibilities. I also became acutely aware of how gendered this form of walking is. With the majority of parental leave still taken by women in the UK, despite the changes entitling shared paternity leave, it is mainly women who walk with babies and young children and who have no doubt had similar experiences to my own. However, these feelings of being encumbered when walking and the inequalities emerging from these experiences are mainly overlooked in both policy and academic discourse surrounding walkability.

In contrast, during my return to work following the birth of my eldest son, I was encouraged to get a bike by my partner as a means of making the journey from home to nursery and then to the office quicker and more efficient. I initially resisted, as I not only found the thought of cycling with a 1-year-old on the back of my bike a daunting prospect, but I couldn't face the cognitive effort of working out the new logistics in terms of clothing, bike storage, helmets, etc. My partner persisted and I eventually got a bike and fitted a child's seat on the back. To say this revolutionised my everyday mobilities is an understatement. I found the whole experience utterly transformational and loved the sense of freedom cycling gave me. In many ways, I think my experiences of cycling emerge as a product of the feelings I experienced as a new mother endlessly walking carrying the physical and emotional burden of caring for a new baby. However, there is another aspect of my experiences of everyday walking that I also consider to be of significance. As many women will attest, women pedestrians are often subjected to a torrent of unwanted attention and sexual harassment. Yet, as a cyclist, there is a sense in which one can move through potentially threatening spaces more quickly, decreasing the chances of unwanted attention. As a woman and mother in my 40s my exposure to such instances has inevitably decreased but I am struck by how my previous experiences continue to feed into my everyday mobility practices. While it might seem strange to be writing about cycling when this is a book about walking, it is relevant to the story I wish to tell here about the significance of the inclusions and exclusions that emerge from our everyday walking practices. My hope is that this book makes visible the importance of understandings of urban walking which place concerns with the everyday politics of pedestrian practices such as these at the centre of analysis.

Acknowledgements

This book has taken a long time to write. I signed the contract when I was expecting my second child, who is now five! However, I do believe that the finished product has benefitted from this longer gestation period and is a different book to the one I would have written five years ago. Like many authors, I have found the writing process a contradictory one, with it being simultaneously a pleasure and a chore. However, I have been extremely fortunate to have had the support of numerous colleagues along the way, many of whom I feel privileged to be able to call friends.

For the last seven years I have been fortunate to be based in the Transport Studies Unit (TSU) at the University of Oxford. One could not ask for a more supportive and collegial space to work. I am extremely grateful for the wonderful colleagues I have encountered there. I am indebted to Tim Schwanen for his immense support, both professionally and personally, and the generosity of his academic engagement with my work. I am incredibly grateful to David Banister who read earlier drafts of the book and provided such insightful and constructive feedback. I cannot thank Debbie Hopkins enough for not only reading earlier drafts with such care and thought but also for her friendship during what have been some personally challenging times. The friendship and professional support of Kirsty Ray and Ersilia Verlinghieri has also been a constant presence over the last few years that I have valued immensely.

The TSU sits within the School of Geography and the Environment, where I have benefitted from the academic stimulation and support of many other wonderful past and present colleagues both there and at Kellogg College. In particular, I am incredibly appreciative of supportive dialogues, often over lunch, with Nihan Akyelken, Idalina Baptista, Robin Cohen, Marion Ernwein, Beth Greenhough, Ian Klinke, Anna Lora-Wainwright, Jo Waters, and Niall Winters. Gillian Rose has been an inspiring mentor and I will be forever grateful for her thoughtful guidance and encouragement. My thanks also to Rich Holden and Gillian Willis for their unflagging and patient administrative guidance and support.

Over the years I have also benefited from encounters and friendships with other scholars who have interests in mobilities, cities, and allied topics, who have encouraged and supported my work in many ways. My heartfelt thanks to Richard Baxter, David Bissell, Steve Brown, Tim Cresswell, Mark Davidson (for being so critical of the mobilities agenda and pushing me to think more carefully about the politics of walking), Sarah Dyer, James Esson, Russell Hitchings, Eleanor Jupp, Loretta Lees, Susan Moore, Scott Rodgers, Justin Spinney, Emma Street, Jon Shaw, and Kevin Ward. Some of the empirical studies in this book have resulted from collaborations with other researchers including Hari Byles, Jon May, Farhan Samanani, and Richard Yarwood. I am thankful for their intellectual contributions and practical support in the conduct of this work.

The research upon which this book is based has been funded by the Economic and Social Research Council (Ref: PTA-033-2003-00014 and PTA-026-27-1500); University of Oxford John Fell Fund; and the Wellcome Trust (Ref: 201583/Z/16/Z). This research would have not been possible without the organisations I have collaborated with and research participants who have been so generous with their time. Many thanks to John Harris, Lidija Mavra, Donnington Doorsteps, Oxfordshire Breastfeeding Support, the Sock Mob, Street Pastors, and the Royal Society for Blind Children. Faye Leerink at Taylor and Francis has shown incredible patience through this extended process and I cannot thank her enough! Thanks also to Ruth Anderson and Nonita Saha for their editorial assistance through the whole publication process; Madeleine Collinge for copy editing the manuscript; Gaurav Dubey for his help with the references; and Ailsa Allen for the map artwork.

Brenda Beard was my A-Level geography teacher and was responsible for convincing me to do a geography degree as opposed to English literature. I will be forever grateful to her for demonstrating to me that there was more to geography than throwing a quadrat around in a sand dune! During my three years as a geography undergraduate at the University of Wales, Swansea (as it was then known), I experienced some of the happiest times in my life. Thanks to Keith Halfacree and Gareth Jones for widening my horizons further as to what human geography could be and for introducing me to what 'critical thinking' actually was. Many of the friendships from those Swansea days have fortunately stood the test of time. I cannot thank the 'Swansea Girls' enough for their encouragement, love, and laughter.

I am also eternally grateful for the love and kindness of my sister Laura and brother-in-law Chris. Gabriel and Etienne, you are a constant reminder of 'overlapping walks of life'. There are no words to describe the joy you bring to me each and every day. My parents, David and Sue, have been unfaltering in their love and support through the highs and lows of the rollercoaster that constitutes both contemporary academic life and beyond. This book is dedicated to them.

Parts of this book draw upon and develop material which has been previously published. Chapter 2, 'Pedestrian infrastructures', includes some material that originally appeared in "'I'm on autopilot, I just follow the route": exploring the habits, routines, and decision-making practices of everyday urban mobilities', in *Environment and Planning A*, 43(12), 2857–2877. An earlier version of Chapter 4, 'Walking rhythms and the politics of time', appeared as 'Stepping in time: walking, time, and space in the city', in *Environment and Planning A*, 41(8), 1943–1961. Chapter 5, 'Pedestrian politics of (non)encounters', includes material, co-authored with Richard Yarwood, which was first published in '"Christians, out here?": Encountering Street-Pastors in the post-secular spaces of the UK's night-time economy', in *Urban Studies*, 52(3), 501–516. In each instance, the original publication has been re-written, re-structured, and further developed for the present book. I am grateful to the original publishers and co-authors for giving their permission to use the material here.

1 Introducing the 'Walkable City'

'You're the book that I have opened. And now I've got to know much more'. Shara Nelson, lead vocalist of British trip hop group Massive Attack, sings these lines as she walks along a section of West Pico Boulevard in Los Angeles in the accompanying video to the critically acclaimed song 'Unfinished Sympathy'. The video was filmed in one continuous shot on Steadicam and follows Nelson purposely walking along the sidewalk, seemingly unaware of her surroundings. The camera follows her disengagement as she strides past a group of men surrounding an aggressive dog, a young boy taking aim with a toy gun, a white couple having an argument, a black father and child holding hands, and a disabled man with no legs zigzagging on a skateboard behind her. She ignores a homeless person pushing a trolley laden with belongings, a woman handing out flyers, and the street vendor selling fruit and vegetables. Nelson crosses an intersection, focusing on the route in front of her. A collision with an oncoming motorbike to her left is narrowly avoided as she moves forward past an older woman, walking at a visibly slower pace to Nelson, who is then pushed past by a teenage couple.

This music video can be read as exemplifying the blasé indifference of the urban dweller (see Simmel, 1971) in relation to many aspects of contemporary urban life, including gentrification, street violence, homelessness, and urban poverty. However, it is also a telling example of how people appropriate space on foot and the richness of what unfolds through the 'simple act of walking'. The video is not only a visual representation of everyday walking and the intersections of gender, class, ethnicity, age, and disability but also highlights central themes in this book relating to the rhythmicity of the walking body and pedestrian encounters. Yet, how should the complexity of these intersections of gender, class, ethnicity, age, and disability be understood in the context of urban walking? How do we avoid treating all walking as the same? This book seeks to address these questions while attempting to dissolve some of the romanticism that surrounds everyday walking in the city, a romanticism that positions walking as an accessible, democratic, inclusive, and emancipatory urban practice. A romanticism that extolls the positive virtues of walking across academic research and the arenas of policy and practice. The central concern driving the development of this book is the everyday politics of urban walking. In other words, the inclusions and exclusions that emerge from everyday experiences of navigating the city on foot.

In December 2019, Channel 4 broadcast a *Dispatches* documentary that provided a devastating window into the lives of over four million children living in poverty in the UK. One of the families featured in the documentary were eight-year-old Courtney, her six-year-old brother, and their mother. They were living in a three-bedroom flat in Cambridge, having fled domestic abuse, yet were having to move again to another part of the country as they could no longer afford the rent due to a bedroom tax[1] being levied on a third bedroom. Of the many disturbing scenes throughout this broadcast of children enduring the contemporary equivalent of Dickensian Britain, due in no small part to brutal reductions in welfare provision, there is one that has particular poignance in the context of this book and its concerns with everyday urban walking. Courtney, her younger brother, and mum are shown visiting a local food bank for the first time, located 2.5 miles (4 km)[2] from their home. They are filmed walking there and back in the early evening darkness of winter as the bus fare exceeded their miniscule daily budget due to delays in Universal Credit[3] payments. As I watched these young children struggle to help carry the heavy shopping bags of provisions from the foodbank on this 2.5-mile journey back to their flat, I wondered what difference a well-designed, 'walkable', high quality, 'pedestrian friendly', public realm would have made to their experiences. The conclusion I quickly came to was, very little.[4]

This book examines the everyday politics of urban walking – but perhaps more precisely, it is concerned with the different ways walking is situated in contemporary urban life. In doing so, I emphasise the need to consider experiences of walking beyond a narrow set of concerns relating to a well-rehearsed and familiar set of issues including, but not limited to, the built environment, low-carbon mobilities, air pollution, and fear/safety. I argue that making a city 'walkable' in terms such as these is meaningless without taking account of broader inequalities relating to concerns not only with class and poverty highlighted in the experiences of Courtney detailed above but also gender, race, disability, and ageing.

Framing the 'walkable city'

The relationship between walking and the city is multiple and complex, with engagements ranging from the rational and planned to the poetic and sensual. Concerns with urban walking have a long history while featuring in contemporary debates within both policy and academic arenas. Urban pedestrian movement is closely associated with, among other things, urban planning, sustainable transport, public space, social mixing, and artistic practices. As a result, there are multiple ways in which walking and 'walkability' are conceptualised, framed, and understood in different contexts. Within urban and transport policy and academic debate, a crude distinction can be drawn between walking being commonly viewed as either a form of 'active transport' or a means for how we come to engage with and know urban space. However, all too often analytical perspectives focusing on walking are either divorced from broader political and socio-economic processes or the everyday experiences of walking practices themselves. For, as Bieri (2017) argues: 'Walking as art, walking as research method, and walkable urbanism can all be understood as an attempt to reclaim an everyday practice – walking. It is striking, however, that the predominant forms that emerge from this endeavor are not everyday forms of walking but performative forms of walking' (29). In this book I provide a critical examination of the socially differential nature of pedestrian experiences in relation to the everyday politics of pedestrian practices. With this, the book tends to practices of everyday walking as its central focus as opposed to the more performative walking practices concerned with walking as art, as a research method, or the commodification of walking through urban development processes.

The central argument of *The Walkable City* is that notions of 'walkability' and 'walkable cities' need to take greater account of how urban walking is imagined, planned for, and experienced. This reconceptualisation of urban walking goes beyond simply asking where, how often, and why people walk towards facilitating a greater understanding of the relationship between walking, urban space, and the everyday politics of contemporary urban life. As such, I am motivated by two overlapping questions:

1 How is urban space produced by different walking practices?
2 What kinds of inclusions and exclusions emerge from how a walkable city is understood and practised?

I seek to engage with these questions through the development of an inter-pretive framework in which walkability is not reduced to where/what is walkable and the frequency of pedestrian movement but the ways in which urban space is both consumed and produced through different ways of walking. In other words, the focus of analysis is not the extent to which cities contain walkable spaces but how people walk in and through urban space. I argue that paying closer attention to the 'hows' of walking makes it possible to gain a greater understanding of the social and cultural inclusions and exclusions emerging from pedestrian practices. Methodological concerns are engaged with throughout the book in terms of both a critical examination of methods for researching walking and walkability, and the possibilities and limitations of walking as a methodological tool in, and of, itself.

In what follows, I examine the intersections between walking practices and concerns with geographies of gender, class, ethnicity, age, and disability while situating pedestrian practices in the complex coordination of everyday life. Walking is not a singular practice but takes on many different forms and characteristics. However, the aim of this book is not to make *a priori* distinctions between different 'types' of walking or produce some form of walking typology but to explore how pedestrian movement features in everyday contemporary urban life. I am particularly concerned with the neglected, banal, everyday forms of walking that are often overlooked when the focus is on walking for leisure or as a mode of transport.

The romanticisation of walking is not limited to artistic, literary, and philosophical engagements but stretches through planning approaches, such as New Urbanism (see Chapter 2 for a more detailed discussion), and health perspectives, such as the World Health Organisation 'Steps to health' framework (WHO, 2007). In *The Walkable City*, I argue that the extent to which walking is something to endure rather than enjoy is rarely reflected upon. In attending to this concern, I develop a conceptualisation of walkability as being socially and materially co-produced in relation to unfolding habits, routines, and practices. This conceptualisation of walkability provides an alternative lens to dominant perspectives in urban and transport research, planning, and practice, for critically engaging with a range of issues relating to the material, embodied, affectual, political, and social dimensions of walking. I argue that this theoretical position enables us to gain a greater understanding of how socially differentiated urban space is produced and consumed on foot. My starting point for stressing the importance of how walking is socially differentiated is to engage with more obvious forms of identity politics, particularly in relation to the ways in which walking can be gendered and racialised (see also Brown and Shortell, 2016). I then move on to examine more subtle forms of differentiation, hierarchies, and privilege that emerge through holding up a magnifying glass to the multiple and complex ways in which people use and appropriate urban space on foot.

Peripatetic ponderings: walking as practice, walking as knowing

There is an ever-increasing policy interest in promoting walking as a low-carbon and healthy form of urban transport with an associated growth of pro-walking organisations. Yet, much of the dominant focus on walking and the built environment in these transport and policy research contexts considers pedestrian movement in relation to rational choice and economic demand. These approaches include engagements whereby walking patterns and trends have been forecast and predicted; the implementation of walking policy has been considered; the health benefits of walking as active travel have been promoted; and cost-benefit analysis of pedestrian modes of transport has been conducted.

Much traditional transport research, and urban and transport policy, focuses on how people can be encouraged to adopt more active forms of mobility such as walking and cycling. There is a pre-occupation with quantifying rates of walking (Lee and Talen, 2014; Ogilvie et al., 2004; Tight and Givoni, 2010); examining barriers, including socio-cultural ones, to pedestrian movement (Frank et al., 2006; Sallis et al., 2006); and concerns with the built environment in terms of the optimal urban design features that facilitate pedestrian movement (Ewing and Handy, 2009; Southworth, 2005). Yet, such work frequently lacks any form of in-depth engagement with the lived experiences of walking and is based on a series of assumptions relating to pre-defined journeys with the built environment being the primary influence on a person's decision to walk, with these decisions being made at a specific time and place somehow outside the practice of walking (see for example Sallis et al., 2006). Furthermore, notions of 'the body', beyond a non-disabled, non-differentiated, and non-encumbered pedestrian, are still largely absent from transport, planning, and health discourses on walkability. As such, the complexity of what happens as people walk is often neglected, leading to the heterogeneity of pedestrian practices often being overlooked.

In this book, I develop an analysis that highlights how pedestrian movement is still largely positioned as a homogeneous and largely self-evident means of getting from one place to another with little attention being given to the importance of the processual and experiential dimensions of walking itself. In doing so I propose an understanding of urban walking that is constitutive of so much more than instrumental movements that can be plotted from A to B; whereby the built environment and the materialities associated with walking are just one aspect of a complex entanglement of the spatio-temporal, rhythmic, social, and embodied dimensions of pedestrian practices.

The lived experiences of moving through the city on foot have a long association with a broad range of social, cultural, and philosophical writings. In particular, walking has gained prominence in the context of understanding the city as a site of encounter and interaction. For example, pedestrian movement is situated in relation to concerns with the democratic possibilities of public space (Jacobs, 1972; Sennett, 1970). While for the urban theorist, Simmel (1971), cities were sites of interaction in the context of the blasé attitude and indifference of the urban dweller as a coping strategy for negotiating the sensory overload of the city and the significance of visual social interactions between strangers as they traverse the city streets. Such perspectives have informed understandings of the micro-interactions that occur when walking in the city (see for example Jensen, 2006).

As de Certeau (1984) developed his theory of everyday practices, he drew specific attention to the practice of urban walking. De Certeau was concerned with how everyday practices such as walking are tactically performed against the strategies of those in a perceived position of power, such as planners, engineers, and architects. He suggests that pedestrian acts carry away and displace 'the analytical, coherent proper meanings of urbanism' (102), whereby walking is an everyday act of resistance. Subsequent studies have specifically focused on the everyday politics of walking through engaging with the work of de Certeau (see for example Edensor, 2000; Morris, 2004), others are concerned with the politicised nature of walking as an artistic practice (see Pinder, 2001, 2011), while classic ethno-methodological texts examine the practical accomplishment of walking together (Ryave and Schenkein, 1974; Wolff, 1973) and philosophers have examined the centrality of walking to the work of prominent male thinkers including Nietzsche, Rimbaud, Kant, Rousseau, and Thoreau (see Gros, 2014).

However, many engagements situated within social and cultural theoretical writings can often be abstract, exclusionary, lack empirical detail, and, as I have suggested already, tend to romanticise walking practices. This, in turn, serves to overlook the more 'mundane' dimensions of pedestrian movement, its significance for how people negotiate urban space in their everyday lives, and the implications for how forms of low-carbon/healthy mobility such as walking are promoted within urban policy. Furthermore, there is a pressing need for more systematic empirical engagements with how people's everyday pedestrian experiences are socially differentiated in order to challenge the assumptions underpinning engagements with walking across much academic writing, and policy and practice, of a white, male, non-disabled, walking body.

There is also a long tradition of performative and artistic engagements with urban walking whereby walking is a medium for various conceptual and performance art interventions (see Butler, 2006; Phillips, 2005; Pinder, 2001, 2011; Rendell, 2006). Much of this work is concerned with walking as an embodied and experiential practice evidenced in the artistic practices of those such as Francis Alÿs, Marina Abramovic, Sophie Calle, Janet Cardiff, and Richard Long, alongside the development of what Smith (2014) refers to as 'walking's new movement' of performance art. However, the argument I develop here (see Chapter 6) is that while such artistic engagements are certainly of value for understanding the relationship between pedestrian practices and urban space, these approaches are relatively detached from everyday experiences of negotiating the city on foot. There are notable exceptions (see for example Arora (2020) on performative walking as an artistic response to the politics of walking in the city in relation to sexualised violence in India and South Africa; Johnson (2019) and their adoption of public walks to increase understanding of the racialised exclusion caused by redlining; and Murali (2017) on decolonising walking in the context of performative practices), yet many performance art pieces engaging with walking emerge from a position of class, race, and/or gendered privilege. I argue the importance of acknowledging the power relations and privilege co-produced through different types of walking while attending to the inequalities emerging from such practices, which can be rendered invisible through many artistic engagements with moving on foot.

Walking has also been frequently drawn upon as a means for engaging with and understanding urban space (Rossiter and Gibson, 2003; Solnit,

2001). The concept of the flâneur (Baudelaire, 1964; Benjamin, 1983; Debord, 1967) as a 'method' for reading city spaces and the recent growth of psycho-geography in the mainstream media (see Self and Steadman, 2007; Sinclair, 2003, 2006) are attempts to map the ambience and 'softer' dimensions of urban life through playful walks that seek to engage with the overlooked spaces and people in the city. There are now well-established critiques of the exclusionary figure of the flâneur as a white, non-disabled, male (see Elkin, 2017; Heddon and Turner, 2010; Murali, 2017; Wilson, 1992). Rather than positioning these forms of pedestrian engagements, such as flâneurie, as an appropriate means for understanding the relationship between everyday walking and urban space, I suggest that what they highlight is the need for a greater sensitivity to the experiential dimensions of pedestrian movement and encounters. In other words, how walking the streets can be drawn upon to study the city's everyday rituals and habits or to emphasise the sensory sensual dimensions of urban life.

The 'new mobilities paradigm' (Sheller and Urry, 2006), or 'mobility turn', is now well established across the social sciences, with increasing attention being paid to the mobility of people and goods, ideas, and information. Watts and Urry (2008) contend that researching mobilities 'as a wide-ranging category of connection, distance and motion transforms social science and its research methods' (862). A perceived need for new methods has resulted in a proliferation of interest in mobile methodologies (see for example Büscher et al., 2011; Fincham et al., 2010), such as walking 'go along' approaches (Kusenbach, 2003) and walking methodologies (Bates and Rhys-Taylor, 2017), yet the uptake of these methods has not always been met with an equivalent level of critical reflection (see Merriman, 2014). In what follows I demonstrate how walking as a methodological approach can be practically applied while maintaining a critical distance from the methodological superiority frequently associated with mobile methods as somehow uncovering more 'authentic' access to experience.

Thinking through walkability: an interpretative framework

Walking is a differentiated mobile practice that creates all forms of inclu-sions and exclusions, yet the way in which everyday walking practices are conceptualised has consequences for how these forms of exclusions are understood and redressed. It is therefore important to consider what is meant by 'urban walking', 'walkability', or 'walkable cities' as this has broader implications for engaging with the politics emerging from pedes-trian practices. This book is indebted to several theorists whose work has informed the unfolding analysis presented here. Rather than attempt to summarise the vast contributions of their work, the coming chapters draw upon, and are influenced by, particular aspects of their writings. Through a critical engagement with scholars including Bergson (1911/1998), de Certeau (1984), Latour (1999, 2005), Massey (2001, 2005), Sacks (1992), and Simmel

(1971), the spatio-temporal, embodied, social, and political dimensions of walking emerge, which are key to engaging with, understanding, and conceptualising 'walkability' and the 'walkable city' more broadly. The conceptual contribution emerging from this form of engagement relates to not only understanding pedestrian practices in the context of how everyday urban mobilities are socially and materially co-produced but also in disrupting the self-evident way in which walking is often positioned within transport studies, health research, planning, and policy.

In particular, the works of Doreen Massey and Bruno Latour have been formative in the development of my ideas relating to how urban walking is understood. Attending lectures by both Massey and Latour during my postgraduate studies left a lasting impression on me as my work on walking progressed. Their eloquent and impassioned voices have always echoed through my head while reading their work. Massey's ideas around the importance of how space is imagined, its dynamic nature, and the ways in which it is relationally produced are deeply embedded in the understanding of urban walking I present here. My emphasis on the significance of the 'stuff' of walking, ranging from the built environment to the shoes we wear on our feet, is indebted to Latour's placing of the social and technical on the same explanatory place. For Latour (2005) anything that alters a 'state of affairs' is an 'actant', whether human or non-human. He contends that this is only not the case if you can maintain that: 'walking in the street with or without clothes ... are exactly the same activities, that the introduction of these mundane implements change "nothing important" to the realization of the tasks' (71). The work of both thinkers has provided me with the theoretical and conceptual tools to consider closely the politics of pedestrian practices in relation to who walks in city spaces, where, and how. In other words, how examining the spatial politics of urban walking practices contributes to understandings of the micropolitics of everyday practices and contemporary urban life.

Book structure

In this book I examine contemporary walkability across five thematic chapters concerned with both the planning for, and practice of, walking. The chapters that follow build discussion around research I have conducted singly or in collaboration with other researchers, including Hari Byles, Jon May, Farhan Samanani, and Richard Yarwood. The arguments developed within, and across, these chapters draw upon a series of empirical examples from this research. The result is a combined analysis of original empirical material across several UK cities including London.[5] This UK focus is complemented with reflections on a range of international secondary empirical and theoretical sources that, in different contexts, engage with urban walking. Although each thematic chapter recognises the significance of walking as a mode of transport, my analysis moves beyond the often taken-for-granted engagements with pedestrian movement within the fields of transport studies and policy. Through examining the lived experiences of everyday walking, I illustrate how and why pedestrian movement in the city needs to be situated in relation to a much broader set of concerns beyond 'pedestrian friendly' built environments, low-carbon mobilities, and forms of active travel. In doing so, *The Walkable City* examines the significance of the everyday politics of urban walking to contemporary urban life.

The book includes some previously published material (see Acknowledgements section for details) that I elaborate upon in order to extend my argument throughout each chapter. The empirical work I discuss is primarily based on the analysis of in-depth interviews, participant diaries, and my own ethnographic fieldnotes.[6] However, there are several points in the book where I have drawn upon auto-ethnographical accounts to deepen my analysis. Auto-ethnography situates the personal experiences and self-reflections of the researcher into the context of their wider research concerns (see Ellis and Bochner, 2000; Hammer, 2013; Lingsom, 2012). This has been a significant part of the process of writing this book. As a white, 'middle class', cis, non-disabled woman, I am aware that my reflections originate from a position of privilege. However, reflecting upon these experiences has helped me to situate some of the main arguments I make, particularly around the socially differentiated nature of walking and the ways in which it continues to be romanticised across academia, policy, and practice.

Chapter 2, 'Pedestrian infrastructures', pays specific attention to pedestrian infrastructures in relation to the broader transport planning and policy context of urban walking through an in-depth exploration of the increasing policy interest in promoting walking as an active form of urban transport, the growth of pro-walking organisations, and the dominant focus on walking and the built environment. The analysis I develop in this chapter problematises the dominance of certain approaches for measuring walkability that are based on the assumption of a causal relationship between urban form and travel behaviour. Instead I argue for the adoption of a relational understanding of pedestrian infrastructures where urban design features and material forms are understood as being co-produced through everyday practices. Through a detailed examination of in-depth empirical material from research on everyday walking in London, I challenge conventional understandings of wayfinding on foot as pre-formed decisions informed by rational intention. In contrast, I consider the ways in which journeys into, and through, the city unfold on foot and the ongoing reconfiguration of these journeys. I also draw attention to the significance of habituated bodies to pedestrian infrastructures and the uneven and distributed nature of habits on foot. This chapter serves as a platform for a more in-depth engagement with the conceptual concerns that are developed through the book for exploring the heterogeneity, differentiated, and overlapping dimensions of walking and its relationship with urban space.

Theorists, such as de Certeau (1984), have long stressed the politicised nature of pedestrian movement, namely how walking practices can be understood as a mode of political resistance, particularly against the regulatory domination of urban planning. However, far less attention has been paid to the politics associated with the type of everyday urban pedestrian practices highlighted in Chapter 2. For example, the inequalities emerging from the differential nature of walking the city receive little attention in urban planning contexts in the promotion of low-carbon mobilities. In Chapter 3, 'Walking as social differentiation', I develop an understanding of politics in relation to the inclusions and exclusions that emerge from different walking practices and the relationship with urban space. I draw upon a range of secondary material in making the argument that how people appropriate space on foot matters to understandings of how urban walking is socially differentiated. In particular, I focus on the complex inter-relations of class, race/ethnicity, age, disability, and the gendered politics of walking. This discussion is structured through the overlapping themes of walking imaginaries, street life, and walking as protest and addresses questions including: Who is included and who is excluded from different visions of walkable spaces in cities? In what ways do different forms of urban walking exclude certain groups? How do different conceptions of walking produce particular forms of inclusionary and exclusionary politics?

In Chapter 4, 'Walking rhythms and the politics of time', I engage with the temporal rhythms of walking while considering the temporal politics associated with the differential nature of walking practices. In encouraging people to use active forms of transport, walking is promoted within transport, health, and urban policy as a quick and efficient mode of transport. In line with much transport policy, walking is often framed in these terms in relation to travel time being considered as 'dead time' people seek to minimise. In this chapter, I build upon work that has sought to challenge dominant linear conceptualisations of time in transport planning and policy as nothing more than clock time passing. The productivity of travel time has been examined in various settings (see for example Jain, 2011, on coaches; Watts and Urry, 2008, on trains), including my own work exploring how walking opens up the possibility of doing things in a way that other forms of urban transport are unable to do (Middleton, 2009). However, while there are other studies that engage with the multiple spatio-temporal rhythms of walking (see for example Matos Wunderlich, 2008) and walking practices as a form of political resistance (de Certeau, 1984), far less attention has been paid to the conflicts and vulnerabilities of pedestrian rhythms in the context of concerns with active forms of mobility. As such, this chapter also draws upon in-depth research on everyday urban walking in London in engaging with experiential time, or people's sense of time, as a means of exploring the multiple temporalities that emerge out, and shape, people's everyday experiences on foot. I draw upon the writings of Adam (1990) and Bergson (1911/1998) to engage with the experience of time and how people become aware of their own duration. A series of temporal inequalities become evident from this focus that make visible the differential nature of everyday pedestrian practices. I conclude by arguing that considering the rhythms of the walking body is essential for further understanding these differentiated pedestrian experiences and associated inequalities of everyday urban walking.

Chapter 5, 'Pedestrian politics of (non)encounters', begins from the starting contention that walking has been positioned as an inherently social activity across a wide range of social and cultural theory, urban, planning, and transport literature. Recent years have also witnessed a growing interest in urban encounters. Underpinning this work is recognition that, despite growing multiplicity and inequality, a certain 'low level sociability' is still apparent in our cities through which people learn to live with difference (see for example Amin, 2006; Butcher, 2009; Thrift, 2005; Vincent et al., 2018; Wilson, 2011). However, while Valentine (2008) welcomes the more positive focus of work on notions of the 'good city', she raises an important note of caution in suggesting that much of this work 'celebrates the potential for the forging of new ... ways of living together with difference *but* without actually spelling out how this is being, or might be, achieved in practice' (324). For example, in the context of walking there still remains very little explicit attention to the socialites of moving through the city on foot.

Chapter 5 attends to such an impasse and draws upon in-depth empirical work on two organisations where pedestrian urban patrolling is central to their activities. First, the Street Pastors, who are an inter-denominational church initiative aimed at providing city centre care for users and providers of the night-time economy. Second, the SockMob, a volunteer network, who participate in regular walks in central London. SockMob are named after the socks they use as an ice breaker for conversations with homeless people they encounter on the streets. The discussion focuses on mobile ethnography, which was used to gain an in-depth understanding of both groups' patrolling activities whereby the significance became evident of the unfolding of their pedestrian practices and associated encounters and non-encounters. In doing so I draw upon the invitation of Kärrholm et al. (2017) to reflect on how different sorts of walking practices are assembled to co-exist (or not) in order to gain new insights into the subtle power relations of urban mobility in public space. Furthermore, through the analysis of in-depth interview and participant observation data emerges a counter position to the frequent romanticised discourses surrounding everyday walking practices that has wider resonance to how we understand the everyday politics of urban mobility in public space. I conclude with a call for a clearer engagement with the socialities of urban walking while raising important questions concerning the politics of both pedestrian encounters and non-encounters. I conclude that walking is not an innocent urban practice and that the power relations emerging from the everyday encounters that do, and do not, unfold on foot matter to how walking in the city is understood.

In Chapter 6, 'Walking bodies, emotional geographies, and mobile methods', I examine the embodied, material, and technological relationality of walking to further disrupt self-evident notions of walkability. I use Judith Butler's question of 'what can a body do?' (Taylor et al., 2010) as a starting point for developing an analysis that focuses on the significance of the embodied and experiential dimensions of urban pedestrian practices in order to counter universalising conceptualisations of walking, which ignore the significance of differentiated mobile bodies. In line with the previous chapters, I situate and understand everyday walking in the unfolding experiences of urban pedestrian practices. Through focusing on how these practices are socially and materially co-produced, I draw specific attention to the embodied, material, and technological relations and their significance for engaging with everyday urban movements on foot. In considering the emotional and affective geographies of pedestrian practices, I attend to a neglected but significant dimension of everyday urban walking.

I also discuss in-depth empirical research on the everyday mobility experiences of visually impaired (VI) young people in London in the broader context of critical reflections on using mobile methods. I argue that the experiences of this group as they navigated the city while engaging with mobile methods, including video elicitation approaches, raise important questions around the accessibility of mobile methods to differentiated mobile bodies. The mobility knowledge and practices emerging in the everyday lives of VI young people, and how these are intimately linked to bodily senses and the materiality of the city, further disrupt the self-evident nature of walking and notions of walkability that feature so prominently in urban and transport research, planning, and practice.

In the concluding chapter (Chapter 7, 'Stepping forwards: some concluding thoughts') I bring together a series of reflections informed by the arguments developed throughout the book, which include a need for:

- increased understanding of the heterogeneous, overlapping, and socially differentiated dimensions of walking and its relationship with urban space;
- focused analytical perspectives on walking in relation to broader political and socio-economic processes and the everyday experiences of walking practices themselves;
- greater sensitivity to the experiential dimensions of pedestrian movement and the disconnect between policy/practice and social/cultural theory.

I draw together my concluding thoughts on the significance of conceptualising walkability as being socially and materially co-produced in relation to the themes that cut across each chapter of walking infrastructures, understanding walking as socially differentiated, and walking methods. I reflect upon the contributions of the book in terms of thinking differently about walking and interrogating its taken-for-grantedness as an everyday mobile practice. In particular, I emphasise the need to decolonise walking knowledges as an important part of challenging the universalising tendencies of 'Western'[7] paradigms, methods, themes, and transport solutions. I stress the significance of how each chapter has made visible:

- the inclusions and exclusions emerging from how everyday urban walking is both imagined and practised;
- the importance of centring the lived experiences of the walking body in the analysis of pedestrian practices;
- the need to take seriously the 'right to mobility' in terms of who and how people appropriate space on foot in understandings of contemporary urban life.

In what follows, each chapter not only reveals a further set of issues essential for comprehending urban pedestrian practices, but also makes visible their inter-relations and interdependencies. My hope is that reading this book provides an opportunity to have not only 'thought about walking' (Solnit, 2001: 16), but also to examine what walkers take to be significant about their urban experiences on foot, and in so doing bring into focus the everyday politics of the 'walkable city' in terms of 'overlapping walks of life'.

Notes

1. The Bedroom Tax is an under-occupancy charge, which was introduced in 2012 as part of a series of brutal austerity policies that included the British Welfare Reform Act. Under the tax, social housing tenants are charged an increased rent if their housing is considered to be surplus to their requirements.
2. To put this into context, 1.3–3 km is considered 'reasonable walking distance' in active transport to school literature and policy. While 'walkable distance' measures are problematic on multiple levels, they are still prevalent and underscore what many people think is acceptable to expect of children (see for example Chillón et al., 2015).
3. Universal Credit was introduced in the UK in 2018 with the aim of merging all benefits, such as housing and income support, into one payment. However, the system has been marred by inefficiencies, particularly around the length of time it takes claimants to receive payments.
4. In a UK study on motherhood and caring for children in poverty, Bostock (2001) found that a reliance on walking as a mode of transport can have a negative impact on the welfare of families.
5. These cities have been anonymised to protect the identity of research participants in the organisations I worked with.
6. Throughout the book, pseudonyms are used for the names of all research participants to protect anonymity.
7. Here I use 'Western' to refer to a set of discourses and philosophies of thought originating from places including the United States, Western Europe, and Australia.

References

Adam, B. 1990. *Time and social theory*, Cambridge, Polity.

Amin, A. 2006. The good city. *Urban Studies, 43*, 1009–1023.

Arora, S. 2020. Walk in India and South Africa: Notes towards a decolonial and trans-national feminist politics. *South African Theatre Journal, 33*, 14–33.

Bates, C. & Rhys-Taylor, A. 2017. *Walking through social research*, London, Routledge.

Baudelaire, C. (translated and edited by Jonathan Mayne). 1964. *The painter of modern life, and other essays*, Oxford, Phaidon.

Benjamin, W. 1983. *Der Flâneur, Das Passagen—Werk, 2 Bde*, Frankfurt, Suhrkamp, 524–569.

Bergson, H. (translated by Arthur Mitchell). 1911/1998. *Creative evolution*, Mineola, Dover.

Bieri, A. H. 2017. Walking in the capitalist city: On the socio-economic origins of walkable urbanism. In: Hall, C. M., Ram, Y. & Shoval, N. (eds.) *The Routledge international handbook of walking*, Abingdon, Routledge, 27–36.

Bostock, L. 2001. Pathways of disadvantage? Walking as a mode of transport among low-income mothers. *Health & Social Care in the Community, 9*, 11–18.

Brown, E. & Shortell, T. 2016. *Walking in the European city: Quotidian mobility and urban ethnography*, London, Routledge.

Büscher, M., Urry, J. & Witchger, K. 2011. *Mobile methods*, London, Routledge.

Butcher, M. 2009. Ties that bind: The strategic use of transnational relationships in demarcating identity and managing difference. *Journal of Ethnic and Migration Studies, 35*, 1353–1371.

Butler, T. 2006. A walk of art: The potential of the sound walk as practice in cultural geography. *Social & Cultural Geography, 7*, 889–908.

Chillón, P., Panter, J., Corder, K., Jones, A. P. & Van Sluijs, E. M. F. 2015. A longitudinal study of the distance that young people walk to school. *Health and Place, 31*, 133–137.

De Certeau, M. 1984. *The practice of everyday life*, Berkeley, London, University of California Press.

Debord, G. 1967. *La société du spectacle*, Paris, Buchet/Chastel.

Edensor, T. 2000. Walking in the British countryside: Reflexivity, embodied practices and ways to escape. *Body & Society, 6*, 81–106.

Elkin, L. 2017. *Flâneuse: Women walk the city in Paris, New York, Tokyo, Venice and London*, London, Vintage Books.

Ellis, C. & Bochner, A. 2000. Autoethnography, personal narrative, reflexivity: Researcher as subject. In: Denzin, N. K. & Lincoln, Y. S. (eds.) *Handbook of qualitative research* (2nd Ed.), London, Sage, 733–768.

Ewing, R. & Handy, S. 2009. Measuring the unmeasurable: Urban design qualities related to walkability. *Journal of Urban Design, 14*, 65–84.

Fincham, B., McGuinness, M. & Murray, L. 2010. *Mobile methodologies*, Basingstoke, Palgrave Macmillan.

Frank, L. D., Sallis, J. F., Conway, T. L., Chapman, J. E., Saelens, B. E. & Bachman, W. 2006. Many pathways from land use to health: Associations between neighborhood walkability and active transportation, body mass index, and air quality. *Journal of the American Planning Association, 72*, 75–87.

Gros, F. 2014. *A philosophy of walking*, London, Verso.

Hammer, G. 2013. "This is the anthropologist, and she is sighted": Ethnographic research with blind women. *Disability Studies Quarterly, 33*.

Heddon, D. & Turner, C. 2010. Walking women: Interviews with artists on the move. *Performance Research, 15*, 14–22.

Jacobs, J. 1972. *The death and life of great American cities*, Harmondsworth, Penguin.

Jain, J. 2011. The classy coach commute. *Journal of Transport Geography, 19*, 1017–1022.

Jensen, Ole, B. 2006. 'Facework', flow and the city: Simmel, Goffman, and mobility in the contemporary city. *Mobilities, 1*, 143–165.

Johnson, W. 2019. Walking Brooklyn's Redline: A journey through the geography of race. *Journal of Public Pedagogies, 4*, 209–216.

Kärrholm, M., Johansson, M., Lindelöw, D. & Ferreira, I. A. 2017. Interseriality and different sorts of walking: Suggestions for a relational approach to urban walking. *Mobilities, 12*, 20–35.

Kusenbach, M. 2003. Street phenomenology: The go-along as ethnographic research tool. *Ethnography, 4*, 455–485.

Latour, B. 1999. *Pandora's hope: Essays on the reality of science studies*, Cambridge, MA, Harvard University Press.

_____. 2005. *Reassembling the social: An introduction to actor-network-theory*, Oxford, Oxford University Press.

Lee, S. & Talen, E. 2014. Measuring walkability: A note on auditing methods. *Journal of Urban Design, 19*, 368–388.

Lingsom, S. 2012. Public space and impairment: An introspective case study of disabling and enabling experiences. *Scandinavian Journal of Disability Research, 14*, 327–339.

Massey, D. 2005. *For space*, London, Sage.

Massey, D. B. 2001. Living in Wythenshawe. In: Borden, I., Kerr, J., Rendell, J. & Pivaro, A. (eds.) *The unknown city: Contesting architecture and social space*, Cambridge, MA, London, MIT.

Matos Wunderlich, F. 2008. Walking and rhythmicity: Sensing urban space. *Journal of Urban Design, 13*, 125–139.

Merriman, P. 2014. Rethinking mobile methods. *Mobilities, 9*, 167–187.

Middleton, J. 2009. 'Stepping in time': Walking, time, and space in the city. *Environment and Planning A, 41*, 1943–1961.

Morris, B. 2004. What we talk about when we talk about 'walking in the city'. *Cultural Studies, 18*, 675–697.

Murali, S. 2017. A manifesto to decolonise walking: Approximate steps. *Performance Research: On Proximity, 22*, 85–88.

Ogilvie, D., Egan, M., Hamilton, V. & Petticrew, M. 2004. Promoting walking and cycling as an alternative to using cars: Systematic review. *British Medical Journal, 329*, 763.

Phillips, A. 2005. Cultural geographies in practice: Walking and looking. *Cultural Geographies, 12*, 507–513.

Pinder, D. 2001. Ghostly footsteps: Voices, memories and walks in the city. *Ecumene (continues as Cultural Geographies), 8*, 1–19.

_____. 2011. Errant paths: The poetics and politics of walking. *Environment and Planning D: Society and Space, 29*, 672–692.

Rendell, J. 2006. Walking. *Art and architecture: A place between*, London, IB Tauris, 181–190.

Rossiter, B. & Gibson, K. 2003. Walking and performing "the city": A Melbourne chronicle. In: Bridge, G. & Watson, S. (eds.) *A companion to the city*, Oxford, Blackwell Publishing, 437–447.

Ryave, A. L. & Schenkein, J. N. 1974. Notes on the art of walking. *Ethnomethodology*, *265*, 274.

Sacks, H. 1992. *Lectures on conversation vol. 1 and II*, Oxford, Blackwell.

Sallis, J. F., Cervero, R. B., Ascher, W., Henderson, K. A., Kraft, M. K. & Kerr, J. 2006. An ecological approach to creating active living communities. *Annual Review of Public Health*, *27*, 297.

Self, W. & Steadman, R. 2007. *Psychogeography*, London, Bloomsbury Publishing.

Sennett, R. 1970. *The uses of disorder: Personal identity & city life*, New York, Knopf.

Sheller, M. & Urry, J. 2006. The new mobilities paradigm. *Environment and Planning A*, *38*, 207–226.

Simmel, G. 1971. *The metropolis and mental life*, Chicago, University of Chicago Press.

Sinclair, I. 2003. *London orbital: A walk around the M25*, London, Penguin Books.

_____. 2006. *Edge of the Orison*, London, Penguin Books.

Smith, P. 2014. Performative walking in zombie towns. *Studies in Theatre and Performance: Zombies & Performance*, *34*, 219–228.

Solnit, R. 2001. *Wanderlust: A history of walking*, London, Verso.

Southworth, M. 2005. Designing the walkable city. *Journal of Urban Planning and Development*, *131*, 246.

Taylor, A., et al. 2010. *Examined life*. [Montréal], National Film Board of Canada [Available at: https://www.tarshi.net/inplainspeak/voices-when-sunaura-taylor-and-judith-butler-go-for-a-walk/].

Thrift, N. 2005. But malice aforethought: Cities and the natural history of hatred. *Transactions of the Institute of British Geographers*, *30*, 133–150.

Tight, M. & Givoni, M. 2010. The role of walking and cycling in advancing healthy and sustainable urban areas. *Built Environment*, *36*, 385–390.

Valentine, G. 2008. Living with difference: Reflections on geographies of encounter. *Progress in Human Geography*, *32*, 323–337.

Vincent, C., Neal, S. & Iqbal, H. 2018. *Friendship and diversity: Class, ethnicity and social relationships in the city*, London, Springer.

Watts, L. & Urry, J. 2008. Moving methods, travelling times. *Environment and Planning D: Society and Space*, *26*, 860–874.

Wilson, E. 1992. The invisible flâneur. *New Left Review*, *191*, 90–110.

Wilson, H. F. 2011. Passing propinquities in the multicultural city: The everyday encounters of bus passengering. *Environment and planning A*, *43*, 634–649.

Wolff, M. 1973. Notes on the behaviour of pedestrians. In: Birenbaum, A. & Sagarin, E. (eds.) *People in places: The sociology of the familiar*, New York, Praeger Publishers, 35–48.

World Health Organization (WHO) 2007. *Steps to health: A European framework to promote physical activity for health*, Copenhagen, WHO Regional Office for Europe.

2   Pedestrian infrastructures

Walking in cities is frequently positioned as the panacea to many of the ills of urban life. According to much urban and transport policy, 'pedestrian friendly' streets provide a range of benefits that include reducing traffic congestion and pollution, encouraging people to exercise, boosting the local economy, and promoting social interaction (Department for Transport, 2003, 2004, 2018, 2020; Department of the Environment, Transport and the Regions, 2000;[1] European Commission, 2004; Gehl, 2011; Transport for London, 2004, 2005, 2018). An overarching theme of this book is the relationship between walking and the ways in which cities are imagined and designed. In this chapter I draw upon vignettes from how walking has been positioned and promoted in London as a lens through which to consider walking in relation to broader concerns with the built environment and urban wayfinding. In particular, I examine how walking as a practice is assembled across the spheres of urban policy and practice while emphasising the significance of the social and material co-production of walking practices for engaging with the complexities of the walkable city.

Policy responses and debates on urban walking are predominantly focused on concerns with the construction of 'pedestrian friendly' physical infrastructures. Here, I contend that while there is much attention given to the materialities of pedestrian infrastructures in planning and designing for urban walking, the lived experiences of these infrastructures are frequently overlooked. As such, walkers themselves are largely absent beyond being seen as homogenised, undifferentiated, and abstract bodies. This chapter attends to such omissions in proposing a relational conceptualisation of 'pedestrian friendly' infrastructures where material forms and everyday practices are understood as co-producing such infrastructures. This conceptual move increases understandings of the complexity of everyday urban walking by situating urban form, in both material and virtual manifestations, in relation to the lived experiences of urban pedestrians. This provides the foundations for subsequent chapters that extend concerns with urban walking to take greater account of the differentiated nature of our spatio-temporal, embodied, and social experiences on foot.

I start by introducing a relational conceptualisation of pedestrian infrastructures. This highlights the significance of both the materialities and everyday practices to the emergence of what are considered 'pro walking' urban design features. I then draw upon this approach to consider notions of 'walkability' in terms of its widespread adoption in pedestrian planning and practice while problematising the dominance of particular approaches for measuring it. The chapter moves on to examine urban wayfinding on foot in light of a relational understanding of pedestrian infrastructures. In doing so I challenge the foundational assumption within much transport research, policy, and practice that if people have access to the right information, use the right technology platforms, in the right environment, they will be encouraged to walk. In contrast, I show how wayfinding is not a static context in which information and technology lands but a set of continually moving elements that are difficult to predict. Through the interrogation of empirical material drawn from in-depth research on everyday urban walking in London, I consider the ways in which journeys into, and through, the city unfold on foot and the ongoing reconfiguration of these journeys beyond an analysis informed by 'rational' intention.

The chapter then engages with Bissell's (2015) contention that greater consideration needs to be given to how infrastructures are affected by habit. I argue that habituated bodies are a coping mechanism for people as they navigate the many challenges of traversing the city on foot. Such habits are not only situated in the unfolding of pedestrian practices but also need to be understood as infrastructures that are unevenly distributed and nonconstant entities. I conclude by stressing the importance of attending to the absence of walking bodies through much work engaging with urban walking across research, planning, and practice. Each remaining chapter of the book seeks to attend to this absence through an in-depth consideration of the spatio-temporal, embodied, and social dimensions of the experiences of everyday walking in contemporary urban life.

Pedestrian 'friendly' infrastructures

Typically, notions of pedestrian infrastructures conjure up certain material images of pavements, walkways, signage, barriers, etc. Yet, these materialities do not exist independently of the people who use them. Pedestrian infrastructures are shaped not just by their material form but through the practices of urban walkers. The concept of 'infrastructure' has become increasingly central across social sciences analysis, in urban contexts in particular, as a way of rendering visible the technical, material, and political systems which can support or constrain everyday lives (Berlant, 2016; Larkin, 2013; Star, 1999). As Schwanen and Nixon (2019) note: 'One of the elements that holds this body of work together is a rejection of common sense notions of infrastructure as "hardware", that is, roads, canals, cables, pipes and so forth' (147). Instead, a relational understanding of infrastructure is process orientated whereby, rather than existing as independent material forms, infrastructures emerge and come into being through everyday practices (Star and Ruhleder, 1996).

The 'infrastructural turn' has its roots in anthropology (see for example Star, 1999) but has been engaged with across a range of critical urban scholarship. At the centre of such work lie interrogations of the 'invisible' and 'taken for granted' (Graham and Marvin, 2001) nature of the 'stuff' of cities. Examples of work adopting such an approach include studies concerned with the urbanisation of water (Swyngedouw, 2004); the enchantment of infrastructure in road building (Harvey and Knox, 2012); cycling infrastructures (Latham and Wood, 2015); and the politics of infrastructure (Larkin, 2013; McFarlane and Rutherford, 2008). Cass et al. (2018) examine how infrastructures intersect with societal transformations, in contexts including cycling in Copenhagen, on-street charging for electric vehicles in Oxford, and food storage in Hanoi and Bangkok. They stress the significance of infrastructures not simply taking material and institutional forms but 'arrangements that are embedded within and constitutive of what people do' (165). As such, their interest lies in both the ways in which infrastructures shape each other and enable certain practices. Conceptualising pedestrian infrastructures beyond materialities serves to highlight a set of subtler processes which attune us to the multiple ways in which everyday walking practices are both materially *and* socially co-produced. I will now consider some of the implications of this thinking for understanding everyday urban walking by focusing first on notions of walkability followed by pedestrian wayfinding and the routines and habits emerging from these practices.

Walkability

The perceived 'walkability' of a neighbourhood or city has long been central to somewhere being labelled as walkable. Yet, what does 'walkability' actually mean and how is it typically understood? The concept has been interpreted and adopted in a multitude of different ways (Forsyth, 2015). For example, Ram and Hall (2017) contend that 'walkability is achieved when the streets and other walking spaces provide pedestrians a secure network of connections to varied destinations, within a reasonable amount of time and effort, and offering a pleasant and interesting context (Southworth, 2005)' (311). In considering walkable places for tourists, as opposed to local residents, they produce a detailed summary of the different dimensions of walkability and the myriad different ways in which it is measured. A combination of objective (such as the built environment and physical geography of a place) and subjective (such as micro-decisions of individuals in relation to routes and distance) factors are discussed as relevant to how walkable a place is.

Measuring levels of walking in relation to infrastructure characteristics and perceived qualities of different spaces dominate many discussions concerned with walkability and the promotion of urban walking across policy and practice. Such measurements are strongly linked to features of the built environment such as pavement/sidewalk quality, car-free zones, street lighting, and the removal of street clutter. Research and practice on walkability is driven by well-meaning intentions to create 'liveable' and 'sustainable' urban futures. Yet the question remains as to why much of this work is so heavily focused on the built environment, physical infrastructures, and homogenised walking bodies while relying predominantly on numerical measures and quantification. Andrews et al. (2012) highlight the exclusionary nature of much work on walkability. They note how walkability is part of the neoliberal welfare state in relation to individuals being encouraged to take responsibility for their own well-being through healthy lifestyles and participate in civic life through their use of public spaces. Their work proposes a need for future research on 'different forms of embodiment, urban mobility and health (particularly regarding disability, frailty and illness)' (1928).

Numerous walkability measures and toolkits have been developed by transport scholars (Adkins et al., 2012; Anciaes et al., 2017; Duncan et al., 2016; Lee and Talen, 2014; Moura et al., 2017; Ujang and Muslim, 2014). The adoption of such measures has been particularly prominent in US city contexts with two measures dominating public understandings of walkability in US urban neighbourhoods. Walk Score began in 2007 with an aim of promoting walkable neighbourhoods. Users could input any address in the United States and receive a score from one to 100 on 'walkability'. This venture was soon seized upon by real estate companies in their marketing activities and became privately owned by a real estate brokerage in 2014. The National Walkability Index was launched by the US Environmental Protection Agency as a public alternative for ranking residential blocks according to their 'relative walkability'.

Quantifying and measuring walkable urban neighbourhoods in such ways is at the heart of the New Urbanist planning movement that emerged in North America in the 1980s as an antidote to automobile dominated, residential suburban sprawl. Uniform developments consisting of houses built in neotraditional architectural styles began to appear in many US, Canadian, and European cities throughout the 1990s and early 2000s. The fundamental principles of the New Urbanist movement concern what are positioned as 'liveable' spaces at a human scale. Planning regulations see the enforcement of on-street parking and the location of 'commercial and civic centres at a walkable distance from most homes, and zone activity spaces for mixed – rather than single – use purposes' (Al-Hindi and Till, 2001: 191).

In a series of blog posts for the Congress for the New Urbanism, Robert Steuteville muses about the effectiveness and accuracy of walkability scores. In stressing the demand for living in mixed-use and walkable communities and the importance of 'walkability' for real estate investment he argues that:

> For researchers, the current indexes muddy the difference between places that are walkable, and places that are not. Good science is based on objective reality. When the National Walkability Index gives a higher score to places where very few people walk compared to places where almost everybody walks, there's a problem.
>
> A better walkability measure would take into account how many people actually walk, compared to how many drive, in a given location. That could be accomplished using a ratio of walking to driving on public rights of way. Walking inside of a building—like a shopping mall, would not count. The only important variable would be how many people walk, relative to how many people drive, to the mall (or wherever).
>
> (Steuteville, 2019)

The concerns raised by Steuteville are framed through a positivist lens of 'good science' being based on 'objective' research. In adopting such a position, the only logical conclusion he can reach is that 'the only important variable' in understanding the walkability of different neighbourhoods is counting 'how many people walk'. In practice, the adoption of walkability scores and measures is much more diverse, nuanced, and complex and can be useful tools in creating inclusive and accessible public realms. For example, Giles-Corti et al. (2015) developed a 'Walkability Index Tool' as a means to address the gap between the need and creation of walkable neighbourhoods. This was informed by aspects of the built environment including street connectivity, residential density, and land use mix. In Yoshii's (2016) research on walkability in historical Japanese cities, concerns with safety, health, and social cohesion emerged as prominent issues (see also Mateo-Babiano, 2016, on pedestrian safety concerns in Manilla). Meanwhile Mitullah et al. (2019) have sought to understand walking patterns and walkability in Kenyan slum communities. They focused on the informal settlement of Mukura Nwa Njenga in Nairobi to conduct a walking and accessibility audit as a means of: (1) understanding pedestrian mobility patterns in the settlement; (2) exploring residents' attitudes to and perception of the walking environment; (3) determining the links between activity spaces and walkability; and (4) understanding and benchmarking the mobility needs of residents in informal settlements.

Embedded within many measures of walkability and assessing how 'walkable' somewhere is are the presumed economic benefits of pedestrian activity. For example, much policy discourse foregrounds and promotes the presumed economic benefits of walking with widespread claims across urban and transport policy of how increasing pedestrian footfall boosts the local economy. This links more broadly to concerns relating to the growth of automobility and out-of-town shopping malls and the gradual demise of city centre retail and the local high street (see Freund and Martin, 2008; Parlette and Cowen, 2011). Scores, indexes, and indicators of how 'walkable' different parts of the city are perceived to be are also strongly related to property prices. For example, in a study of Washington DC, Alfonzo and Leinberger (2012) suggest that developers, investors, and local and regional planning agencies should consider walkability in their decision-making in relation to property investments (see also Pivo and Fisher, 2011). While it is perfectly feasible for there to be a relationship between levels of walking and the economic development of an area, it is one that is problematic and should be treated with caution. Bieri (2017) critically refers to this as the 'walkability fix' in relation to the commodification of walking and how walkable urbanism is 'a sales pitch for real-estate planning' (34). This 'sales pitch' and 'spatial form of capitalism' (34) with social consequences which are exclusionary to certain groups is a far cry from the emancipatory potential often linked to everyday walking.

Although it is possible to identify value in walking scales, indexes, and measures, concerns with appropriate data on walking in terms of pedestrian environments, activities, and behaviours have also featured prominently in policy and practice arenas over the years (Brög and Erl, 2001; Desyllas et al., 2003; Gemzoe, 2001). For example, in 2006, the pro-walking organisation Walk21 launched a project to standardise methods for measuring walking. This resulted in the publication of the International Walking Standard in 2016 where a series of key performance indicators were identified as relevant for measuring walking and as a means of developing 'reliable, valid and yet easy to use travel surveys for cities and urban areas which include walking in a consistent and appropriate manner' (Sauter et al., 2016: 4). These indicators include average number of daily trips made, travel time, distance travelled, and mode share. Yet, other considerations need to be taken into account in measuring walking such as scale, heterogeneity, and the quality of the data.

Nearly 20 years ago Gemzoe (2001) warned, in the context of research on the pedestrianisation of Copenhagen, that 'one of the key factors in understanding the complexity of areas for walking is that there is much more to walking than walking ... Numbers alone are not an indication of the quality of a place' (20). There are a series of assumptions relating to the accessibility of walking embedded in measurements and scores that walking is something that is accessible to all as long as there are sufficient 'pedestrian friendly' environmental features such as quality sidewalks and pedestrian crossings. A central aim through the forthcoming chapters of this book is attending to this problematic assumption and that we need to pay greater attention to the socially differentiated nature of walking (see Chapter 3). In particular, attention is drawn to the significance of different people's experiences of everyday walking while illustrating the importance of recognising that not all walking is the same. Unlike Steuteville, I do not discount certain types of walking but instead seek to attend to the neglected experiences of urban walking, those which are either overlooked by policy discourse or do not conform to urban imaginaries of what constitutes walking in the city, including the mundane, uncomfortable, and unspectacular aspects of everyday pedestrian practices.

Measuring walking has become a prominent feature of many people's everyday lives through the advent of our increased interactions with communication technologies such as smartphones and watches. This rise in the use of technologies that can measure, chart, assess, and make publicly available our everyday mobilities coincides with public health campaigns highlighting the importance of walking in tackling growing health concerns with sedentarist lifestyles (see Gov.UK, 2018; Public Health Agency, 2020). Walkability is central to debates within public health on the significance of active travel in the creation of 'healthy cities'. Concerns about an 'obesity epidemic' are well rehearsed and frequently drawn upon in the promotion of forms of active travel such as walking and cycling (Lopez and Hynes, 2006; Ogilvie and Hamlet, 2005; Patterson et al., 2017; Warin et al., 2008). Within this discourse explicit links are made between the built environment and its effects on physical activity. The obesogenic thesis (Hill and Peters, 1998) is a prominent feature of such work where low-cost food high in sugar and saturated fat combined with low levels of physical activity in car-dominated environments are considered the key factors in increasing levels of obesity in adults and children (Schwanen, 2016). The work of Lawrence Frank and colleagues (Frank et al., 2005; Saelens et al., 2003) has been particularly influential here with Knaap and Talen (2005) going as far as stating that Frank and Engelke (2005) 'confirm our suspicions that human health is related to urban form and that greater attention to urban form can increase physical activity and perhaps improve human health' (110).

The public health agenda on walking and its focus on walkability measures have particular implications for how bodies are imagined. Not only are walking bodies frequently assumed to all be the same with little account taken of bodily differences, but there is a strong rhetoric underpinning much of this work relating to the disciplining of bodies. In other words, all bodies needing to conform to a particular imaginary of a mobile body that is typically male, non-disabled, and not 'overweight' or 'obese' (see Colls and Evans, 2014; Guthman, 2013, on problematising the obesogenic thesis). Dudley Shotwell (2016) refers to this as 'healthism', where health is understood as an individual and moral obligation. In the context of walking, Springgay and Truman (2019) highlight how the lens of healthism makes visible 'the ways walking can be commodified, convey moral judgements, and exclude particular bodies from access and mobility' (3).

Furthermore, embedded within this emphasis on the built environment is a set of causal relationships between 'pedestrian infrastructures' and 'pedestrian behaviours', with the former impacting upon the latter. In other words, urban walkers are passive subjects whose mobilities are shaped and dominated by urban form. These are problematic assumptions as little or no account is taken of how everyday urban walking practices also shape pedestrian infrastructures. Adopting a relational understanding of pedestrian infrastructures enables the co-production of their emergence through *both* materialities and practices to be explored. The association between levels of walking and specific urban forms emerging from the New Urbanist agenda and work within public health, form a powerful discourse influencing the thinking that informs urban, transport, and pedestrian policy (see also work of Gehl, 2011, on public space).

Since Transport for London (TfL) launched *The Walking Plan for London* in 2004 and its bid to be one of the most 'walking friendly cities' by 2015 (changing to 2020 in 2010 and then to '80 per cent of all trips in London to be made on foot, by cycle or using public transport by 2041' in 2018), the message has been clear: get the built environment and the quality of information right and more people will be encouraged to walk. In 2020, TfL published a *Planning for Walking Toolkit* with the aim of providing planners and designers with advice in the planning and design of 'good environments for walking'. While the handbook should be applauded for acknowledging the complexity of walking, in ways that previous documents had failed to do, and the needs of different pedestrians (disabled, older people, young people, etc.), there are still assumptions embedded about the walking body. For example, in addressing what is termed an 'inactivity crisis', there is considerable discussion of the notion of 'comfort' including 'human comfort assessments'. Comfort is understood here as relating to the adequate allocation of space for walking, pedestrian flow, and unhindered movement. Yet it is important to note that comfort for some does not necessarily equate to comfort for others and that we cannot assume that continuous, unhindered pedestrian circulation is desirable for all (for example, see Chapter 3 for discussion on informal street vending in Hanoi and Accra). Relatedly, such toolkits are of course also based on the premise that all people welcome the opportunity to walk. To attend to this over-emphasis on urban form, and a move away from causal relationships between the built environment and pedestrian trips, requires re-conceptualising 'pedestrian infrastructures' as relational processes that are materially and socially co-produced. I will now turn to the example of shared space to elaborate this point further.

Shared space is a form of street design whereby the physical barriers separating motorists and pedestrians are removed as a means of traffic calming. Shared space is an urban design concept visible in many European cities and is frequently positioned by its advocates as a pro-walking design feature that encourages walkability. The underpinning philosophies of shared space are strongly linked to New Urbanist thinking and its emphasis on pedestrianised routes due to roads being considered as a significant barrier to social interactions and the creation of liveable streets (see for example Appleyard, 1980). Increased interactions between different road users are central to the concept of shared space with a view to creating more convivial urban spaces. For example, Jonasson (2004) provides an empirical examination of a traffic intersection in Gothenburg as a means of understanding the micro-interactions between pedestrians, cyclists, motorists, etc. The notion of shared space can also be situated in the context of such concerns in relation to negotiations between different road users.

However, as with New Urbanism, concerns were raised in relation to the increasing support and implementation of shared space schemes across the UK. For example, Imrie (2012) questioned the extent to which shared space is a suitable design feature of the built environment for more vulnerable road users, such as those who are visually impaired. He argues that shared space is an 'auto-disabling' environment and has the potential to compromise the safety and well-being of these groups through increasing contact with motorised forms of transport. Imrie's contention is that shared space 'serves to reproduce auto-dominated environments that are creating, potentially, new spaces of disablement in towns and cities, particularly for vulnerable pedestrians, such as the elderly and vision-impaired people' (2261). Such views have also been strongly echoed by disabled transport activists as they point to the dangers and risks of shared space for disabled and neuro-diverse people (see Disabled Persons Transport Advisory Committee, 2018). This has led, following the publishing of an *Inclusive Mobility Strategy* by the DfT, to the pause in development of shared space schemes in the UK as guidance is reviewed. However, the legacy of the ideas, visions, and imaginaries of shared space schemes remain.

One of the most notable shared space schemes in London is Exhibition Road in South Kensington, which was implemented in 2012. The street forms a corridor between South Kensington tube station and Hyde Park with several significant museums and academic institutions located there including the Natural History Museum, the Science Museum, Imperial College, and the Royal Geographical Society (RGS). Since 2004, I have regularly attended the RGS-IBG Annual Conference, which is held in both the RGS building and Imperial College. Like many delegates I have made the regular walks up and down Exhibition Road between the two conference venues. I never paid too much attention to that short walk, too engrossed in conversations with colleagues and friends or rushing to the next session, until 2016 when I attended with my five-month-old baby. I was shocked by the sudden vulnerability. I experienced as I tried to cross a section of road where black cabs hurtled around the corner where a roundabout once stood. The following year I walked with my eldest son, who was four years old at the time, towards the RGS and tried to unsuccessfully explain how and when it was safe to cross. This was not an easy task in a space that felt anything but shared.

If we adopt a conceptualisation of pedestrian infrastructures that is concerned with how infrastructures shape each other and enable certain practices, it is not simply the materiality of the 'pedestrian friendly' urban design feature of shared space that is significant but the practices which shape its emergence. In other words, the design features of shared space (such as removal of drop curbs and designated crossings) need to be understood in relation to how people are using it, with no clear priority for pedestrians and coaches and vans obstructing access on foot. From this perspective, shared space only comes into being from what people do, with its success as a 'pedestrian friendly' infrastructure being clearly dependent on an entanglement of materialities, practices, and histories. While residents in South Kensington might welcome the improved aesthetics of the area, the historical legacy of motorised transport, especially tourist coaches, dominating the area continues. For, as Cass et al. (2018) argue, a relational understanding of infrastructure 'highlights the extent to which infrastructures and practices intersect in ways that are shaped by their combined and separate histories. These intersections have consequences for future intervention, for what policy makers and others can do, and for how their actions (past and present) affect social, material and political arrangements, and for how these interact, now and in years to come' (166). As such, how people move emerges as a significant dimension that requires more sustained attention if we are to further examine how walking is both materially and socially co-produced. This conceptual move serves to highlight how changes to the physical infrastructure alone, such as dropping curbs and removing street clutter, will not change the mobility cultures of a particular place, redress inequalities, or democratise everyday mobilities and transport modes. In contrast, in what follows, I draw upon a relational understanding of pedestrian infrastructures in focusing on wayfinding on foot, and the associated everyday decisions to walk, as a means of engaging with these concerns.

Wayfinding

I moved to London in 1999 and for much of the 10 years I lived there I, like many others, relied heavily on a well-thumbed A–Z to navigate my way around unfamiliar parts of the city. Yet on the occasions that RMT[2] industrial action stopped London Underground services, I was always struck by the sight of commuters, who had travelled into London by Overground train, popping up like lost moles blinking in the bright light while trying to find their way to their offices in an unfamiliar maze of city streets. This was a pre-smartphone era where an A–Z was the dominant navigation tool. In similar terms to most forms of travel, having the correct wayfinding information has long been associated with facilitating pedestrian movement with policymakers and practitioners frequently launching different schemes based on this assumption. For example, the TfL-funded Legible London Project began in 2007 with the aim of promoting and making walking easier through a range of consistent information including street signage and maps (see Figure 2.1). The scheme ran on a trial basis in the Bond

Street area but is now in most of London's 33 boroughs with many similar schemes appearing in other cities across the UK. The aim of the coordinated street signs is to show pedestrians where to walk, how long it will take, and significant landmarks along the way with TfL (2008) arguing that Legible London in the Bond Street area has sped up the average walking

Figure 2.1 Legible London scheme

Source: Image: Author's own

journey by 16%. Like much transport analysis, speed is understood as key, with shorter travel times a central benefit (see Tranter and Tolley, 2020, on speed, cities, and sustainable transport). It should be noted that for some people, temporal measurements of how long it takes to get somewhere are easier to understand than how far somewhere is measured by distance. Yet these 'KPIs' of the scheme assume that if people have access to the right information, in the right environment, they will then decide to walk. In what follows I challenge this assumption in relation to everyday wayfinding practices on foot.

Around the same time as the Legible London scheme was launched, Islington Council produced 'Walk Islington: Explore the unexpected' (see Figure 2.2). The walking guide detailed six walking routes through the borough with 'tips to motivate you to get to know your borough better' (Islington Council, 2007). According to the Senior Transport Planner who designed the guide, it was aimed at 'young professional women' who were considered (by Islington Council research) to be the group most likely to adjust their travel behaviour to walking.

Figure 2.2 'Walk Islington: Explore the unexpected', Islington Council walking guide

Source: Copyright © 2007 Composite Projects

However, as she explained:

> One of the major barriers to them [women] walking more was their lack of knowledge of the local walking environment, they were afraid to walk into unknown areas and didn't know the routes through their local area. I therefore sought to create a guide that was targeted to women, and that would give them motivation to walk more – identifying routes through the borough and places of interest along the way.
>
> (Interview – Senior Transport Planner,
> Islington Council, June 2007)

These 'places of interest along the way' included cafés, shops, and galleries within a design that was considered 'female sensitive'. Five thousand copies of the walking guide were distributed around local shops, libraries, and leisure facilities and it was available to download from the Council website. The gendered nature of the guide is striking. This is not simply in relation to its design incorporating flowers, jewellery, and a coiffured mannequin head with long eyelashes and painted red lips or the implicit assumptions that places of most interest to women include shops and cafés. Hanson (2010) argues that gender and mobility are 'completely bound up with each other, to the point of almost being inseparable' (6). Feminist geographers and transport scholars have long emphasised how women typically experience shorter but more complicated and local everyday mobilities (Hanson and Hanson, 1981; Hanson and Pratt, 1995; Howe and O'Connor, 1982; Law, 1999). In reflecting upon such work, Uteng and Cresswell (2008) highlight how 'studies establish that gender-differentiated roles related to familial maintenance activities place a greater burden on women relative to men in fulfilling these roles resulting in significant differences in *trip purpose, trip distance, transport mode and other aspects of travel behaviour*' (3; emphasis in original). Women are more likely to walk due to caring responsibilities and having less access to economic resources that would allow them to move by other transport modes (Hoai Anh and Schlyter, 2010). Murray (2020) highlights how the walking commute is gendered and generational with women, children, and older people (intersecting with race, class, ethnicity, disability, and sexuality) continually disadvantaged while moving on foot due to infrastructure planning prioritising the needs of individual motorised journeys that still tend to be undertaken by men (see also Kern, 2020, on the feminist city and Criado Perez, 2019, on the invisibility of women in transport planning). A guide identifying places deemed to be 'of interest' to women was unlikely to address their differential walking experiences. Rather, in seeking to target women in the promotion of walking in the local area, the Islington Council promotional campaign was a representation that both re-enforces and re-produces such differential gendered everyday mobility patterns.

With over 1,800
dedicated pedestrian
street signs, walking
around London is
easier for everyone.

We work with partners across London
to deliver easy-to-use signage

Search TfL Improvements

MAYOR OF LONDON

TRANSPORT
FOR LONDON
EVERY JOURNEY MATTERS

Figure 2.3 Transport for London walking poster

Source: Copyright © 2019 Rose Blake

More recently, TfL walking promotional campaigns continue to be framed around providing accurate information for pedestrians. For example, on a recent tube ride on the Bakerloo Line through Baker Street station, I glanced up to see a poster of a duck being followed by several ducklings with an image of a Legible London information plinth in the background (see Figure 2.3). The text accompanying this image stated how 'with over 1800 ... street signs, walking around London is easier for everyone'. Yet, this prompts the question of 'easier' for whom? Our engagements with wayfinding information are differentiated. It is therefore important to consider how people use wayfinding information and how this is situated in the unfolding of their everyday pedestrian practices. I will subsequently return to this point later in the chapter.

Since walking promotion schemes such as Legible London were launched, technology has moved on, with smartphones completely transforming the ways in which we move through cities in our everyday lives. In his paper 'iPhone city', Bratton (2009) argues that the iPhone, and presumably other smartphones, now overshadows both the car and physical city and that smartphones have resulted in mobility evolving from mechanical to informational. He highlights how smartphones have physically slowed users down due to people orientating to the screen, causing them to largely overlook their physical surroundings. One extreme example of this is a 30-metre 'cell phone lane' in Chongquing, China, for people using their phones while walking. In arguing how smartphones withdraw users from physical surroundings, Bratton also points to how people are closer to the city in terms of real-time data and information. He claims 'the device becomes a window onto the hidden layers of data held in or about the user's immediate environment. Urban and network diagrams are images now animated in hand, transformed from maps into image-instruments with which to connect and control the immediate and remote environment' (Bratton, 2009: 93). In discussing how smartphones can disengage us from physical engagements with the city, Bratton also focuses on the opportunities that smartphones provide for engaging with urban space and contexts through elements such as location-based services and the social features of phones. Google Streetview provides an interesting example of an application that assists with orientation and wayfinding while simultaneously withdrawing the user from, and engaging the user with, their physical surroundings.

More recently, Lyons (2020) has sought to examine walking within the broader context of such concerns and our reliance on digital platforms in facilitating everyday mobility. Via a Tweet on social media in January 2019, he proposed the concept of 'walking as a service' (WaaS). Lyons (2020) argues that walking has been a neglected mode of transport in the growth of Mobility as a Service (MaaS), defined as users having 'access to multiple modes through a single platform, as opposed to having to engage directly with multiple mobility, information and transactional services across different modes' (279). Showing multiple services on one platform in attempts to provide integrated mobility solutions is a relatively recent development with CityMapper and Google Maps being the most established mobile apps in the context of walking. Both apps integrate data from all modes of urban transport. Lyons highlights both the benefits and potential of walking, while attributing relatively low levels to 'the cognitive challenge – when faced with the unfamiliar – of judging how long a journey could take on foot and determining how to navigate to the destination' (271). He proposes Google Maps Navigation as a digital platform solution for providing users with virtual assistance that 'holds your hand' while wayfinding on foot. Lyons argues that a WaaS business model is an alternative to MaaS, which is based on selling access to mobility. MaaS is a business and governance model that has a large presence in Scandinavia and is driven by entrepreneurs, yet, to date, there has been little work critically examining this growth (see Jittrapirom et al., 2017; Smith et al., 2019). In contrast, WaaS is founded upon selling access to geography and consumers.

However, what unites both MaaS and WaaS models is a generic conception of 'the user' with bodies being largely absent and abstract entities. As such, there is little acknowledgement of the differentiated experiences of walking bodies. For example, Google Maps is programmed to assume that pedestrians walk at a minimum speed of 3 mph. An obvious point being that there is clearly variation as to the different speeds that people can walk. The implications of this vary, as sometimes travel time matters and sometimes it does not, with some people adapting this information knowing that it is not suitable for their walking pace and rhythm. A second assumption embedded in such apps, which can be argued to promote and reinforce the legitimacy and importance of the apps, is that walking is an individualised activity. Yet walking is not an individual accomplishment but a practice that is collectively produced. The ways in which people walk together have altered due to how smartphone technologies and associated apps have become central to our wayfinding practices. In Laurier et al.'s (2016) study of walking with map apps they argue that greater attention should shift away from 'the classical lone walker' to the collaborative work people do, and interactions that unfold, as they walk 'together with touchscreens' as navigation tools (132). WaaS is certainly an interesting and, to date, unexplored proposition. Yet it is also informed by the assumption that if people have access to the right information and the right environment, they will make a decision to walk.

In contrast, I now turn to consider the ways in which journeys into, and through, the city unfold on foot and the ongoing reconfiguration of these journeys beyond an analysis informed by 'rational' intention.[3] Accounts such as Lyons's (2020) position technology as a coping mechanism for the challenges of urban walking. In focusing on the habits and routines of urban pedestrian mobilities such as the daily commute, I argue that habitual behaviour can also be framed in such terms. In contrast to transport research that conceives of habit as an external force, existing outside of everyday practices that somehow obstructs more positive sustainable travel behaviour, I examine the significance of how everyday urban walking unfolds and of pedestrian habits as infrastructure. I argue that this form of engagement makes it possible to understand habit as situated and part and parcel of the sequentially organised and occasioned performance of journeys on foot. This, in turn, enables the transformative potential of habitual behaviour to be realised along with its significance to how pedestrian infrastructures emerge and are sustained through everyday practices.

Everyday experiences of urban walking

There are assumptions that inform many understandings of walking across policy, practice, and academic research that walking is a homogeneous and largely self-evident mode of transport. I have argued elsewhere that far too little attention is given to the experiential dimensions of walking (see Middleton, 2010, 2011). This includes addressing questions such as what are the different types, forms, and characteristics of urban walking and what does it mean to different people? It was questions such as these that served as a departure point for research I conducted on everyday urban walking in London. In particular, this work aimed to address a series of issues that included exploring the relationship between walking and built environment; the many different types, forms, and characteristics of walking; and the social dimensions of moving on foot. However, the overarching aim of the research was to focus on the experiential dimensions of urban walking and explore the significance of what actually happens between A and B as opposed to quantifying the frequency of pedestrian activities. The research took place across a transect through the inner London boroughs of Islington and Hackney (see Figure 2.4), cutting through Barnsbury in Islington, then eastwards into Canonbury, through De Beauvoir Town and London Fields in Hackney. This transect was selected because of contrasts within, and across, each area in relation to the built environment and social concerns such as levels of wealth and deprivation (see Butler and Robson, 2003; Butler and Rustin, 1997). This allowed the walking experiences of different groups of people to be explored.

A mixed-methods approach was adopted that included a postal survey, experiential walking photo diaries, and in-depth interviews. The following draws upon a range of diary and interview accounts produced by some of the 36 participants (18–29 years, 14%; 30–39 years, 41%; 40–49 years, 17%; 50–59 years, 14%; over 60, 14%) who detailed their urban walking experiences (see Middleton, 2009, 2010, for further discussion of the research setting and discursive analytic approach). In focusing on pedestrian experiences, issues associated with decision-making and everyday life coordination became prominent concerns. Although dominant understandings of walking across much transport research, policy, and practice are framed around predetermined decisions to walk or not, I argue here that decisions to walk need to be understood not only in relation to the built environment or wayfinding information but also as habitual behaviour that is intimately bound up with people's everyday routines. I draw attention to the way these understandings assist in not only reconceptualising decision-making processes in the context

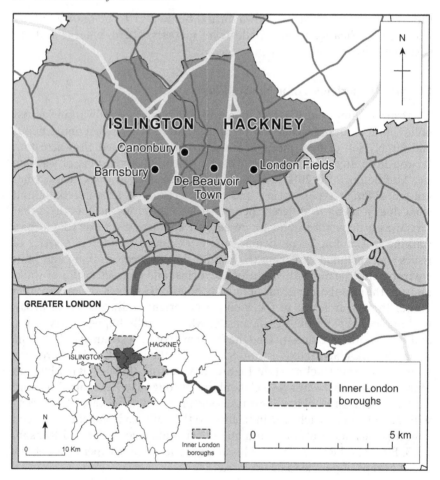

Figure 2.4 Map of transect

Source: Map artwork: Ailsa Allen

of everyday urban mobilities such as walking but also how pedestrian infrastructures are materially and socially co-produced.

Walking routines and household coordination

Following trip to polling station started to make my way to work (again with husband, baby + dog) stopping on the way for a coffee on Chapel Market. I had arranged for Solanta, our nanny, to meet us there. She had a cup of tea with us and took over with the baby and the dog. My husband and I then left for work, him by bike + me by foot'.

(diary – Gemma, Barnsbury resident, Islington)

Gemma's diary extract describes a recent journey to work. What the extract illustrates is that walking is a key resource in her household's morning routine. By the time they leave the coffee shop, part of the journey to work has been completed, the dog has been walked, childcare responsibilities have been handed over to the nanny, and the family have spent some time together. The decision to walk is accounted for in terms of the coordination of their everyday routine, rather than the result of the built environment or wayfinding (see Dowling, 2000; Schwanen, 2011, on feminist perspectives of car use and the role of the car in managing complex daily routines). Dominant understandings of decisions to walk as somehow being taken outside the practice of walking can in part be attributed to much research being theoretically underpinned by models from behavioural psychology such as the Theory of Planned Behaviour (Ajzen, 1991). This approach maintains that human behaviour is the result of individual attitudes and behavioural intentions influenced by social norms and exercising free will. It has been particularly influential in health research (although see Sniehotta et al., 2014, on the contentions surrounding this approach and need to 'retire the theory of planned behaviour'). In a similar vein, ecological models of walking and cycling behaviour position a range of external factors, including social, cultural, organisational, or environmental influences, as impacting upon, as opposed to being situated within, these mobile practices. These models have also been incredibly influential in public health research (see Sallis et al., 2015) and in debates on walking in the context of encouraging active travel (see for example Ginja et al., 2018; Pikora et al., 2003; Saelens et al., 2003).

In the data extract below, Ros reflects upon her journey to work. Her decisions could be analysed through an ecological model in relation to the physical infrastructures and environmental influences upon her journey. However, such an approach overlooks the distributed nature of pedestrian decision-making practices and the complexity of how the journey to work may or may not be accomplished:

> Typical journey [to work] is bicycle door to door unless it's my turn or I have to take my son to nursery in which case it's a combined walk and bike ride. If the weather is good I put him on a bike seat on the back of the bicycle and I walk him because I don't like the traffic, even though I'm a seasoned cyclist I will not cycle him, he's only a year old, on the back. So I put him on the bike seat on the back of the bike, walk him to nursery which is about, round about a 12–15 minute walk straight from the front door and once I've dropped him off I carry on cycling into work and then I reverse that on the way home.
>
> (interview – Ros, De Beauvoir Town resident, Hackney)

The reasons for walking in Ros's account are interpolated not only into the journey to work but also into the contingencies of whether or not it is Ros's turn to take her son to nursery. Walking is made a salient part of their journey in terms of a number of combinations. Her son becomes the bicycle's passenger as she pushes the bike; her journey on such days is a further hybrid of cycling and walking. In making visible those combinations she demonstrates that whether or not to walk not only assists the coordination of her daily routine of journeying to work but is also occasioned within the household's responsibilities in relation to her son's nursery attendance. Sheller and Urry (2006) argue that transport research is misleading, as it tends 'to examine simple categories of travel, such as commuting, leisure, or business as if these were separate and self-contained' (212) (although see Krygsman and Dijst, 2001, on multimodal trips).

Historically, National Travel Survey data on walking in the UK was a good example of this in which a trip was defined as 'a one way course of travel having a single main purpose' and 'walk trips are those where walking is the main mode in terms of distance' (Department for Transport, 2008). However, more recently this has changed in the transport policy context with increasing attention being paid to what people do during trips and a less rigorous separation of trips and stationary activities. As such, in the most recent National Travel Survey, one-way course of travel is defined as a 'trip' while acknowledging that these 'trips' can consist of one or more stages where there is a change in the mode of transport (Department for Transport, 2018). For example, Ros does not have a singular 'purpose of travel' for her 'trip on foot'; rather, her movements involve the dual tasks both of getting her son to nursery and then continuing on to work on time. Following such logic, she is taking two trips of one stage each. Yet Ros's account is also a further reflection of the significance of understanding walking in relation to the practical accomplishment of everyday household coordination activities emphasised by Jarvis et al. (2001). The routines and practices that emerge from and sustain such coordination activities are key elements in the co-production of pedestrian infrastructures.

Ehn and Löfgren (2010) focus on the 'invisibility' of routine as they explore 'the large undercurrent of routines, creating rhythms and temporalities in everyday life, rarely noticed or reflected upon' (99). In the context of train commuters in southern Sweden, O'Dell (2009) argues that it is important to focus not only on the significance of commuting in people's lives but also on broader concerns relating to how we understand the routines in people's everyday lives. The significance of routines to everyday urban walking patterns is rarely engaged with in much transport research, planning, and practice yet I have illustrated the importance of understanding decisions to walk in the context of these types of everyday household routines. However, in what follows I turn attention to how notions of habit and the habitual also need to be taken account of in a relational conceptualisation of pedestrian infrastructures that contributes to better understandings of the complexity of everyday urban walking patterns.

Thinking with habit

Habitual behaviour has increasingly been identified as a key dimension in the promotion of sustainability agendas. One of the most renowned pieces of work influencing these arenas is Thaler and Sunstein's (2009) book entitled *Nudge: Improving decisions about health, wealth, and happiness* (see also Sunstein, 2019, *How change happens*). Nudges are essentially low-cost interventions in product design, infrastructures, and processes, which do not prohibit choice, while seeking to 'defrost' existing habits and lock in new ones. This influential work has gained considerable traction in many spheres of policy and practice,[4] including concerns with the ways in which people travel and how to encourage more sustainable travel behaviour (see for example Avineri, 2012; Lyons, 2004; Metcalfe and Dolan, 2012). For example, Nudge Theory informed much of the thinking of how TfL planned for the increased volume of passengers across London's transport network during the 2012 Olympics. Yet, across these bodies of work, habits have been positioned as 'barriers' to more sustainable behaviour. For example, Prillwitz and Barr (2009) argue that 'habitual behaviour quickens and simplifies acting for a person, but the outcome for others is maybe less beneficial or even disadvantageous than it would have been as a result of a deliberate decision process. For measures trying to influence individual travel decisions, habits are very obstructive' (3). Furthermore, there is a tendency to focus on the individual in relation to habitual behaviour. For as Verplanken et al. (1997) explore the role of habit in the process of information use in daily travel-mode choices, they argue that expectancy value models 'emphasize the deliberate character of individual decision making' (2).

Shove's (2010) work examines approaches embedded in contemporary environmental policy in relation to social change. She refers to the dominant model as 'the paradigm of ABC – attitude, behaviour, and choice' (1273) and in doing so draws attention to the limitations of such an approach to the challenges of climate change. Shove is particularly critical of some of the work of behavioural psychologists, such as Paul Stern and Nudge theories, in relation to the way habit is conceptualised and drawn upon to explain situations that cannot be situated within this ABC model of behaviour:

> For Stern, as for other psychologists, the notions of habit provide a means of importing concepts of context, positioning this as a driver of behaviour in cases where volition and choice are evidently lacking. But if we take a step back, the idea that habits drive behaviour is really odd – implying, as it does, that habit is not itself behaviour but is, rather, some abstract factor bearing down upon the behaviours it directs. (1276)

Instead, Shove argues that habit needs to be situated in the context of social practices and the ways in which practices emerge, change, come together,

and fall apart within complex systems. She stresses the importance of centring analysis on 'how practices evolve, how they capture and lose us, their
carriers, and how systems and complexes of practices form and fragment'
(1279).

There is also a long lineage of philosophical writings engaging with
notions of habit and habitual experience that date back to Aristotle.
One of the most influential thinkers on habit is Ravaisson (2008), whose
work informed the writings of philosophers including Bergson, Deleuze,
and Heidegger. Ravaisson is concerned with the dual logic of habit as
mechanistic repetition but also the freedoms this engenders. His work
blurs, and thus challenges, the mind–body dualisms within the writings of
those such as Kant who find habit problematic. In challenging mechanistic
understandings of habit as restrictive, Ravaisson's work explores the transformative potential of habit in relation to the way repetitive action makes
movements and mobility become easier, which results in greater freedom.
Habit is positioned as dynamic, involving a series of corporeal competen-
cies that refine action over time.

In recent years there has also been a growing body of more empirically
situated work that seeks to explore how habits and routines emerge and are
sustained in everyday life in relation to theoretical notions of social practice. For example, Hitchings (2010) examines the habitual behaviour that
city office workers become engaged with in relation to spending increasing amounts of time in climatically controlled indoor environments, while
Doody (2020) draws upon the philosophies of Dewey in understanding habits as dynamic, relational, and temporal in the context of the everyday (im)
mobilities of New Zealand migrants in London (see also Schwanen et al.,
2012, on Ravaisson and Dewey for re-thinking habits in the context of
decarbonising passenger transport). In a related vein, Bissell (2018) explores
the significance of such conceptualisations of habit for how workers in
Sydney develop the required 'skills' to cope with often arduous commuting
experiences. And Ramsden (2017) drew upon performative art practices as
she invited research participants in Easton, Bristol, UK, and Ann Arbor,
Michigan in the United States, to intentionally interrupt their everyday
habitual pedestrian activities as a means of engaging differently with their
neighbourhoods and surroundings.

In the context of long-duration air travel, Bissell (2015) argues that
there has been much work across many disciplines on the materialities
of mobility infrastructures but there needs to be greater consideration
given to how these infrastructures are affected by habit. He suggests
that 'if habits are apprehended as virtual and distributed, rather than
internal to individual bodies, habit becomes a key part of this [mobility]
infrastructure' (132). He works through a concept of 'virtual infrastructure' in considering how habits change and become configured differently. Drawing upon the example of parking practices and regulations in
England, Merriman (2019) also stresses the significance of engaging with

the distributed nature of habits, practices, and movement in developing an understanding of what he terms 'relational governance'. He argues that such an approach reveals how 'attempts to govern and shape mobility are underpinned by environed understandings of embodied practices, habits and governmental technologies' (1400). In focusing on the role of material objects in strategies to govern and influence the behaviour of individual citizens, he argues, in similar terms to Bissell, for 'more distributive, relational and affective approaches' to understand 'practices, habits, behaviours and events' (1414).

In the last decade, habit has become increasingly important in the types of transport research that predominantly inform urban and transport policy. In this vein of work habit is positioned as an external force somehow driving behaviour, which, as I note above, is, in part, a consequence of conceptions of habit being drawn from particular strands of behavioural and social psychology. As such, much transport research understands habit as something that is located within the individual as opposed to something distributed and relational. In contrast, for the remainder of this chapter I explore the productiveness of understanding everyday urban walking in relation to emergent and processual notions of habit that are both situated within, and co-produced through, pedestrian infrastructures. This conceptual move sits in stark contrast to reductive conceptions of habit as somehow being an external force impacting pedestrian movement and the dominance of approaches that only consider the materialities of pedestrian infrastructures. Analytically, I draw attention to the significance of how people talk about habit and how notions of habit and the habitual are used as resources in the framing of people's accounts of their everyday pedestrian practices. I argue that these accounts not only demonstrate how habits are unevenly distributed (see Bissell, 2015) but also how people become aware of their habitual behaviour only when it is interrupted.

Unfolding journeys, habit, and habitual behaviour

> Yeah, yeah, I always walk, which takes about 25 or 30 minutes. It's so boring! Because it's, I think the quickest route, just because I think it's automatic pilot, I think I just, just walk the route. I even usually walk in the rain but if, for example, you know we had those really icy times, when the streets like literally are an absolute hazard, that's probably the only time I've considered getting on a bus. I think the journey would take me about the same amount of time on a bus. You know I think it would take about half an hour at that time of day and it's just much more unpleasant, you know in really overcrowded public transport. And also the expense I suppose.
>
> (interview – Alice, London Fields resident, Hackney)

This London Fields resident was asked about her typical journey to work and whether she used the same route each day. Her interview extract is just one example of many in which walking can be understood in terms of everyday routines discussed in the previous section. In this instance it is the routine of commuting to work that features in her account. Highmore (2004) is concerned with 'experiential aspects of routine' (307) and states that 'the daily commute to work is perhaps one of the most distinctive modern routines' and that 'commuters often describe their lack of attention to the actuality of these journeys, undertaken it would seem, as if on autopilot' (310). This is certainly a concern in Alice's account of her 'automatic pilot' journey to work on foot. However, more is at stake here in relation to habits that also emerge from everyday routines such as those articulated in Alice's account. Highmore points out, 'there is something ambiguous and problematic about routine ... what routine feels like, how it is experienced, is by no means clear ... because routine is not only dictated from above. We establish our daily routines to give our lives rhythm and predictability' (307). Yet, similar questions can be raised in relation to the habitual. How do routines become habitual, what does habit feel like, how do we become aware of habitual movements on foot, and does a regular walking routine necessarily mean it is habitual? Furthermore, how are habits part of pedestrian infrastructures?

In his work on indoor/outdoor practices and seasonality, Hitchings (2010) explicitly asked research interview respondents about their habitual behaviour. In doing so, he sought to challenge work suggesting it is 'inappropriate' to talk to people about their practices because they will be unable to discuss the habitual (62). But what is interesting in the data I am discussing here is that I did not ask participants directly about their habitual walking behaviour, yet they frequently framed their understandings of everyday urban walking in such terms. In other words, notions of habit are drawn upon to articulate their everyday experiences on foot as opposed to these being shaped and informed by a series of pre-determined decisions. For example, Alice not only makes relevant both practical knowledge relating to her everyday routine, such as knowing the quickest route on foot and the length of the bus journey, but she also demonstrates the relevance to her of forms of 'embodied habits' in terms of walking on 'autopilot' and the hazards associated with the icy streets. Such habits are entangled parts of the pedestrian infrastructures that emerge through her everyday walking practices. Both habit and how people become aware of habit are claimed as experienced in Alice's account. It is through this focus on the discursive organisation of participants' accounts (see Edwards, 1997; Potter, 1997) that we can continue to see the significance of how habit is actually talked about:

> Regardless, I approach Northampton Road. Usually this is an unpleasant part of the walk. Often gangs of kids hang around in the park on either side riding oversized bikes into pedestrians too slow to get out

of the way. One evening, as I passed, I remember a couple of lads setting off fireworks and letting them shoot across the path as pedestrians dodged, ducked, dipped, dived and dodged. I said a large part of
walking is about dodging things. But today those coming the other way
are a couple with a pushchair. These two, coupled with the smell of a
freshly rained upon London make this part pleasant today and I enjoy
the walk. I also enjoy the next thing that happens. A cyclist makes a special effort to stop before a zebra crossing to let me cross. Unexpected,
but welcome.

(diary – Paul, Canonbury resident, Islington)

The fireworks going off on the path, the couple with the pushchair, and the
cyclist who stops can be taken to exemplify this occasioned unfolding of
walking in Paul's account. However, Paul's diary entry is itself organised
in terms of the unfolding and sequentially organised features of his experience, both of his habitual walking patterns (in terms of what he is reminded
of) and in relation to the salience to him of his walking on a particular day.
In addition, his account does far more than identify the habitual nature of
movement he engages in as an urban walker. His accounting is also organised in terms of the unfolding sensuality of walking. The entry is framed in
terms of the contrast between what is 'usually ... an unpleasant part of the
walk' with experiences that remediate his habitual sense of that 'part' in
his walk (the unexpected presence of others and their courtesy, smells, his
enjoyment).

Furthermore, despite taking the same route to and from work each
day and the occasioning of the diary entry in relation to 'typicality' or
his habitual pedestrian experiences ('Usually this is an unpleasant part
of the walk'), his examples also illustrate the way people become aware
of habit when it is interrupted. In his writings on embodiment, habit,
and the everyday, Harrison (2000) draws attention to the work of Varela
(1992) and how the example of a lost wallet is used to illustrate the experiential dimensions of habit. The reader is asked to imagine walking
down the road when the realisation strikes and panic sets in that your
wallet is missing. Harrison points out that there has been some form
of interruption to the sequence: 'relaxed readiness, a breakdown, a gap
opens up, and then an emergent order takes place' (503). He goes on to
suggest 'that it is habit which surrounds such encounters and resolu-
tions, and it is habit which is interrupted in intervals' (503). For Paul,
the pedestrians who are too slow to get out of the way of the fireworks,
walking being about physically 'dodging things', and the cyclist stopping
to let him cross the road are all examples of interruptions that make him
aware of his habitual mobility on foot and reminders of how habits are
part of how pedestrian infrastructures come into being. In similar terms
to how the physical barriers to walking (street clutter, poor pavements,
etc.) typically dominate discussions of urban walking, we can also move

on to ask in what ways habits both support and undermine people negotiating the urban environment on foot.

In writings on pedestrians in the context of automobile growth in the interwar period in London, Hornsey (2010) draws particular attention to government and policy attempts to regulate and automate pedestrian movement.
He highlights that Taylorism was at the heart of instilling urban pedestrians
with correct and automated corporeal habits due to them being considered
'prone to distraction and absent-mindedness' (106) that contributed to a
growing number of accidents and fatalities. Measures such as guardrails
and designated pedestrian crossings were introduced with varying success
rates as a means 'to automate pedestrians and recodify their practices as
a disciplined set of habituated responses' (111). The limited success was in
large part due to the complexity of London's street network coupled with the
need for a more nuanced engagement with predicting walking behaviour.
Hornsey concludes that 'pedestrian behaviour became less about habitu-
ating certain corporeal actions, than about instilling a type of intelligent
responsiveness that could readily adapt to all unforeseen circumstances'
(111). In returning to Paul's diary we can begin to understand not only how
habitual actions emerge from pedestrian practices themselves rather than
something imposed upon urban walkers through the governance of pedestrian movement, but also how habitual behaviour is a means for coping with
the challenges of negotiating the city on foot:

> This time last week I had just come back from holiday and my traffic
> light awareness function was not quite operating at its optimal capacity.
> You get a sense of timing when doing the same route each day. Where to
> cross and lose least time. Which corners are blind and make the heart
> race. And which order the traffic lights change in so you're one pace
> ahead of the infrequent walkers.
>
> Last week my timing was all over the place – now I'm back, a fact
> brought to my attention as Toby runs to follow me across the road.
> Point to note: I never run across the roads. It's just not dignified.
>
> (diary – Paul, Canonbury resident, Islington)

Paul's diary entry draws attention to his 'sense of timing' and the durational
qualities of his journey on foot in relation to the way time expands and
contracts (see Chapter 4 for more detailed discussion of time, space, and
rhythm). He frames this entry around what he terms his 'traffic light awareness function' and how it was not working to its 'optimal capacity' after
he returned from holiday. Paul moves on to qualify the features of this
'awareness function' in relation to knowledge of where to cross the road and
the order of traffic lights in order to increase the efficiency of his movements.
All of this relates to what he positions as a 'sense of timing', which emerges
from the routine of his daily walk to work. Paul emphasises the significance of these dimensions of his pedestrian experiences by contrasting the

pace of his movements on foot with those of 'infrequent walkers'. However, there is more at stake in Paul's diary entry than the speed and efficiency of his routine journey to work. Paul's account also makes relevant the significance of habitual behaviour both for the way we understand decision-making processes in the context of urban walking and how habits are entangled with, and distributed through, pedestrian infrastructures.

Various writings draw attention to the way people cope with the challenges of urban life. For example, Simmel (1971) focuses on the way the body can be conceived of as shutting down to cope with the sensory overload of urban life. In similar terms, Bull (2000, 2005) explores the use of Walkmans and iPods as a means by which people manage everyday urban life through a form of sonic disengagement with their surroundings. More recently, Watson and Drakeford-Allen (2016) highlight how urban dwellers use music to 'tune in' to their environments in ways that enhance their engagements with the city and those within it. Paul's account is one of several examples whereby urban walkers rely on their habitual movements as a strategy for coping with the challenges of negotiating the city on foot. Paul's focus on his 'sense of timing' illustrates that journeys on foot have less to do with a series of pre-conceived decisions taken outside pedestrian practices as positioned in transport policy than they relate to a series of unfolding habitual movements.

Paul's diary entry illustrates that his habitual movements are what enable him to position himself as an accomplished urban pedestrian. He emphasises this accomplishment not only by contrasting it with the 'infrequent walkers' and his friend Toby's 'undignified' dash across the road but also with his own movements the previous week when he had just returned from holiday and how his 'timing was all over the place'. These features of his diary entry illustrate the way walking routines transform over time while also highlighting the significance of maintaining the repetition of movements on foot for walking routines to become habitual. In other words, walking routines are not necessarily habitual routines. This is a significant conceptual point as it emphasises the importance of making an analytic distinction between conceptions of habit and notions of routine.

It also demonstrates the transformative potential of habit as opposed to the restrictive ways in which it is understood within much transport research and policy previously discussed. This transformative potential of Paul's habitual movements enables him to be 'in synch' with his environment. In line with Ravaisson's (2008) focus on the dual logic of habit, movement becomes effortless and receptivity declines, while, simultaneously, spontaneity increases and a space-time emerges which allows a focusing of attention on other concerns. In the context of urban walking, these concerns might include concentrating on maximum efficiency, minimising a perceived loss of time, or getting to the destination as quickly as possible. Nonetheless, as we have seen in participants' accounts, habitual movements are far from enduring and stable. Bissell (2015) highlights that while there has been much work on how habits develop, far less attention has been given to how they can subside and 'decay'. He addresses such concerns through focusing on three transition points in a long-duration airline journey to explore the fragility of habit and how 'the force of habit can wax and wane' (127). Yet, what is the wider significance of a relational conceptualisation of pedestrian infrastructures that centres the material, social, and habitual entanglements that co-produce everyday walking practices, particularly in relation to the ways in which urban walking is imagined and assembled within research, policy, and practice?

Conclusion

Walking is frequently understood to be a homogeneous and largely self-evident means of transport (Middleton, 2010, 2011). Across many spheres of transport research, policy, and practice, walking is something that is planned for and measured with far too little attention given to experiential dimensions. Neglecting what actually happens between A and B results in a lack of understanding of the differentiated nature of everyday pedestrian experiences. In this chapter I have argued the need to focus beyond walkability measures and engage more carefully with how the built environment and wayfinding technologies are understood in relation to encouraging people to walk. In particular, adopting a relational conceptualisation of 'pedestrian friendly' infrastructures, such as shared space, highlights the significance of how both material form and everyday practices co-produce the emergence of these infrastructures. This conceptual shift allows us to move away from dominant understandings of walking whereby walkability is measured, for example, according to the perceived quality of the built environment for pedestrians. The chapter has also critically considered the adoption of wayfinding information and technologies as tools for encouraging people to undertake more everyday journeys on foot. I have shown how it is not simply a case of providing people with information, such as how long it takes to travel somewhere on foot, in order for them to be able to walk. Rather, understanding needs to increase of how wayfinding technologies do not operate in a static context in which information is transmitted and received but are a set of constantly moving elements that are also difficult to predict. Exploring the decision-making processes associated with the everyday urban walking practices of the research participants featured in this chapter has made visible such concerns.

This chapter has also made the argument that it is important to consider how pedestrian infrastructures are affected by habit (see Bissell, 2015). One particular set of insights emerging from this relates to the way habit and habitual behaviour are conceptualised in the context of a relational understanding of pedestrian infrastructures. As opposed to reductive conceptualisations of habitual behaviour, whereby habit is positioned as an obstructive, external force driving individual behaviour, I have drawn attention to how habit needs to be understood as situated in the processual and unfolding action of journeys on foot and as part and parcel of pedestrian infrastructures. This approach sits in contrast to the forms of decision-making and understandings of pedestrian infrastructures that characterise much work across transport research, planning, and practice. In particular, the significance of 'virtual infrastructures' (see Bissell, 2015) has been recognised in how we understand the distributional and relational nature of pedestrian infrastructures. Furthermore, several features of participant's accounts have illustrated how walking routines transform over time and that it is the maintenance of these routines that makes them become habitual. It is possible to have multiple walking routines, yet the movements that unfold from these routines are not necessarily habitual. Therefore, there is an analytic distinction to make between notions of habit and routine.

Although I have acknowledged methodological concerns relating to the inappropriateness of the medium of talk for researching routine and habitual practices, the significance of how habits are actually 'talked about' has also emerged. Despite research participants not being directly asked about their habitual walking behaviour, understandings of everyday urban walking are frequently framed in such terms. Notions of habit are drawn upon as resources for how people frame their everyday experiences on foot. This focus on the discursive organisation of participant's accounts therefore highlights the significance of the medium of talk for how habitual behaviour is researched.

As the data discussed has highlighted, walking in the city can be a challenging experience. Such challenges are often neglected in urban and transport research and practice. The argument I have made here has illustrated the way in which habitual movements are a coping mechanism for pedestrians as they negotiate the city on foot in terms of the ways in which many journeys on foot unfold as if pedestrians were on 'autopilot'. It is these habits that become part of, and emerge out of, the infrastructures that enable walking. This habitual behaviour associated with the everyday practices of urban pedestrians facilitates both a decline in receptivity and an increase in spontaneity that are able to assist the ease with which pedestrians negotiate urban space. A focus on habit, and how walking routines become habitual through the practices of frequent walkers, enables a greater understanding of the nuances of what it is to be an urban walker.

Deleuze (2001) contends that habits are ways of being. We do not have habits that exist independently, instead we are nothing but habits. It is conceiving of habit as situated within the ongoing reconfiguration of journeys on foot that makes visible its transformative potential as opposed to the restrictive associations preventing more 'positive' and sustainable travel behaviour that feature in much transport research. Writings within social and cultural theory have long engaged with how the body moves in these ways (see Lefebvre, 2004; Simmel, 1971). However, this significance needs to also extend to urban and transport research, policy, and practice. For example, an area might be considered more 'walkable' if the flow of pedestrian movement could seamlessly flow on 'autopilot' (Middleton, 2011). It is habitual behaviour that facilitates such movement. Habits can be conceived of as part of everyday mobility infrastructures but it is important that assumptions about the walking body in relation to habit are also interrogated (see Chapter 6). For while habituated bodies are a coping mechanism for walking in the city, habits are inconsistent and fragile entities that are both temporally and spatially unevenly distributed and socially differentiated. It is to the socially differentiated dimensions of walking that I now turn.

Notes

1. Although it is worth noting that the Department for Transport's (DfT) most recent cycling and walking strategy, entitled *Gear Change: a bold vision for cycling and walking*, is dominated by discussions of cycling infrastructures with little mention of walking. This reflects the current Prime Minister Boris Johnson's pre-occupation with cycling dating back to his time in office as Mayor of London. His foreword to the document is framed around 'the joy of cycling' with only passing reference to walking. Cycling is referred to 172 times throughout the document compared to 73 references to walking.
2. The National Union of Rail, Maritime and Transport Workers.
3. The term 'rational' is used here in the context of a narrow economic framing of rationality and instrumentality.
4. See also the work of the Behavioural Insights Team, who draw upon these approaches to 'inform policy, improve public services and deliver results for citizens and society' (Behavioural Insights Team, n.d.). Such behavioural understandings were central to the UK response to the Coronavirus (COVID-19) outbreak in 2020.

References

Adkins, A., Dill, J., Luhr, G. & Neal, M. 2012. Unpacking walkability: Testing the influence of urban design features on perceptions of walking environment attractiveness. *Journal of Urban Design: People in the Design of Urban Places*, 17, 499–510.

Ajzen, I. 1991. The theory of planned behavior. *Organizational Behavior and Human Decision Processes*, 50, 179–211.

Alfonzo, M. & Leinberger, C. B. 2012. *Walk this way: The economic promise of walkable places in metropolitan Washington, DC*, Washington, DC, Brookings Institution.

Al-Hindi, K. F. & Till, K. E. 2001. (Re)placing the new urbanism debates: Toward an interdisciplinary research agenda. *Urban Geography: The New Urbanism and Neotraditional Town Planning*, 22, 189–201.

Anciaes, P. R., Nascimento, J. & Silva, S. 2017. The distribution of walkability in an African city: Praia, Cabo Verde. *Cities*, 67, 9–20.

Andrews, G. J., Hall, E., Evans, B. & Colls, R. 2012. Moving beyond walkability: On the potential of health geography. *Social Science & Medicine, 75*, 1925–1932.

Appleyard, D. 1980. Livable streets: Protected neighborhoods? *The Annals of the American Academy of Political and Social Science, 451*, 106–117.

Avineri, E. 2012. Application area 2: Travel demand modelling: Travel behaviour research. *Transportation Research Circular*. E-C168, November.

Behavioural Insights Team. n.d. Covid 19 – A behavioural perspective [available at: https://www.bi.team].

Berlant, L. 2016. The commons: Infrastructures for troubling times. *Environment and Planning D: Society and Space, 34*, 393–419.

Bieri, H. 2017. Walking in the capitalist city: On the socio-economic origins of walkable urbanism. In: Hall, C. M., Ram, Y. & Shoval, N. (eds.) *The Routledge international handbook of walking*, Abingdon, Routledge, 27–36.

Bissell, D. 2015. Virtual infrastructures of habit: The changing intensities of habit through gracefulness, restlessness and clumsiness. *Cultural Geographies, 22*, 127–146.

_____. 2018. *Transit life: How commuting is transforming our cities*, Cambridge, MA, The MIT Press.

Bratton, B. H. 2009. iPhone City. *Architectural Design, 79*, 90–97.

Brög, W. & Erl, E. 2001. Walking: A neglected mode in transport surveys. Walking the 21st Century, International Conference, Perth, Western Australia.

Bull, M. 2000. *Sounding out the city: Personal stereos and the management of everyday life*, Oxford, New York, Berg.

_____. 2005. No dead air! The iPod and the culture of mobile listening. *Leisure Studies, 24*, 343–355.

Butler, T. & Robson, G. 2003. Negotiating their way in: The middle classes, gentrification and the deployment of capital in a globalising metropolis. *Urban Studies, 40*, 1791–1809.

Butler, T. & Rustin, M. 1997. Rising in the East? The regeneration of East London. *Capital & Class, 21*, 158–160.

Cass, N., Schwanen, T. & Shove, E. 2018. Infrastructures, intersections and societal transformations. *Technological Forecasting & Social Change, 137*, 160–167.

Colls, R. & Evans, B. 2014. Making space for fat bodies?: A critical account of 'the obesogenic environment'. *Progress in Human Geography, 38*, 733–753.

Criado Perez, C. 2019. *Invisible women: Exposing data bias in a world designed for men*, London, Chatto & Windus.

Deleuze, G. (translated by P. Patton). 2001. *Difference and repetition*, London, Continuum.

Desyllas, J., Duxbury, E., Ward, J. & Smith, A. 2003. Pedestrian demand modelling of large cities: An applied example from London. *CASA Working Paper Series* 62.

Department of the Environment, Transport and the Regions. 2000. *Encouraging walking – Advice to local authorities*, London, The Stationary Office.

Department for Transport. 2003. *On the move: By foot – a discussion paper*, London, The Stationary Office.

_____. 2004. *Walking and cycling: An action plan*, London, The Stationary Office.

_____. 2008. *National travel survey: 2007*, London, The Stationary Office.

_____. 2018. *National travel survey: England 2017*, London, The Stationary Office.

_____. 2020. *Gear change: A bold vision for cycling and walking*, London, The Stationary Office.

Disabled Persons Transport Advisory Committee. 2018. Guidance: DPTAC position on 'shared space' [available at: https://www.gov.uk/government/publications/dptacs-position-on-shared-space/dptac-position-on-shared-space].

Doody, B. J. 2020. Becoming 'a Londoner': Migrants' experiences and habits of everyday (im)mobilities over the life course. *Journal of Transport Geography*, 82.

Dowling, R. 2000. Cultures of mothering and car use in suburban Sydney: A preliminary investigation. *Geoforum, 31*, 345–353.

Dudley Shotwell, H. 2016. *Empowering the body: The evolution of self-help in the women's health movement*. Unpublished PhD thesis, The University of North Carolina, Greensboro [available at: https://libres.uncg.edu/ir/uncg/f/DudleyShotwell_uncg_0154D_11873.pdf].

Duncan, D. T., Méline, J., Kestens, Y., Day, K., Elbel, B., Trasande, L. & Chaix, B. 2016. Walk score, transportation mode choice, and walking among French adults: A GPS, accelerometer, and mobility survey study. *International Journal of Environmental Research and Public Health, 13*, 611.

Edwards, D. 1997. *Discourse and cognition*, London, Sage.

Ehn, B. & Löfgren, O. 2010. *The secret world of doing nothing*, Berkeley, London, University of California Press.

European Commission. Environment Directorate-General. 2004. *Reclaiming city streets for people: Chaos or quality of life?* Brussels, Office for Official Publications of the European Communities.

Forsyth, A. 2015. What is a walkable place? The walkability debate in urban design. *Urban Design International, 20*, 274–292.

Frank, L. D. & Engelke, P. 2005. Multiple impacts of the built environment on public health: Walkable places and the exposure to air pollution. *International Regional Science Review, 28*, 193–216.

Frank, L. D., Schmid, T. L., Sallis, J. F., Chapman, J. & Saelens, B. E. 2005. Linking objectively measured physical activity with objectively measured urban form. *American Journal of Preventive Medicine, 28*, 117–125.

Freund, P. & Martin, G. 2008. Fast cars/fast foods: Hyperconsumption and its health and environmental consequences. *Social Theory & Health, 6*, 309.

Gehl, J. 2011. *Life between buildings: Using public space*, London, Island Press.

Gemzoe, L. 2001. Copenhagen on foot: Thirty years of planning & development. *World Transport Policy and Practice, 7*, 19–27.

Giles-Corti, B., Macaulay, G., Middleton, N., Boruff, B., Bull, F., Butterworth, I. & Christian, H. 2015. Developing a research and practice tool to measure walkability: A demonstration project. *Health Promotion Journal of Australia, 25*, 160–166.

Ginja, S., Arnott, B., Namdeo, A. & McColl, E. 2018. Understanding active school travel through the behavioural ecological model. *Health Psychology Review, 12*, 58–74.

Gov.UK. 2018. Focus on brisk walking, not just 10,000 steps, say health experts. Public Health England [available at: https://www.gov.uk/government/news/focus-on-brisk-walking-not-just-10000-steps-say-health-experts].

Graham, S. & Marvin, S. 2001. *Splintering urbanism: Networked infrastructures, technological mobilities and the urban condition*, London, Routledge.

Guthman, J. 2013. Fatuous measures: The artifactual construction of the obesity epidemic. *Critical Public Health: Obesity Discourse and Fat Politics: Research, Critique and Interventions, 23*, 263–273.

Hanson, S. 2010. Gender and mobility: New approaches for informing sustainability. *Gender, Place & Culture, 17*, 5–23.

Hanson, S. & Hanson, P. 1981. The impact of married women's employment on household travel patterns: A Swedish example. *Planning – Policy – Research – Practice, 10*, 165–183.

Hanson, S. & Pratt, G. 1995. *Gender, work, and space*, London, Routledge.

Harrison, P. 2000. Making sense: Embodiment and the sensibilities of the everyday. *Environment and Planning D: Society and Space, 18*, 497–517.

Harvey, P. & Knox, H. 2012. The enchantments of infrastructure. *Mobilities, 7*, 521–536.

Highmore, B. 2004. Homework: Routine, social aesthetics and the ambiguity of everyday life. *Cultural Studies, 18*, 306–327.

Hill, J. O. & Peters, J. C. 1998. Environmental contributions to the obesity epidemic. *Science, 280*, 1371–1374.

Hitchings, R. 2010. People can talk about their practices. *Area, 44*, 61–67.

Hoai Anh, T. & Schlyter, A. 2010. Gender and class in urban transport: The cases of Xian and Hanoi. *Environment & Urbanization, 22*, 139–155.

Hornsey, R. 2010. "He who thinks, in modern traffic, is lost": Automation and the pedestrian rhythms of interwar London. In: Edensor, T. (ed.) *Geographies of rhythm: Nature, place, mobilities and bodies*, Farnham, Ashgate.

Howe, A. & O'Connor, K. 1982. Travel to work and labor force participation of men and women in an Australian metropolitan area. *The Professional Geographer, 34*, 50–64.

Imrie, R. 2012. Auto-disabilities: The case of shared space environments. *Environment and Planning A: Economy and Space, 44*, 2260–2277.

Islington Council. 2007. *Walk Islington: Explore the unexpected*, London, Islington Council.

Jarvis, H., Pratt, A. C. & Wu, P. C. C. 2001. *Secret lives of cities*, London, Prentice Hall.

Jittrapirom, P., Caiati, V., Feneri, A. M., Ebrahimigharehbaghi, S., Alonso-Gonzalez, M. J. & Narayan, J. 2017. Mobility as a service: A critical review of definitions, assessments of schemes, and key challenges. *Urban Planning, 2*, 13.

Jonasson, M. 2004. The performance of improvisation: Traffic practice and the production of space. *ACME: An International E – Journal for Critical Geographies, 3*, 41–62.

Kern, L. 2020. *Feminist City: Claiming space in a man-made world*, London, Verso.

Knaap, G. & Talen, E. 2005. New urbanism and smart growth: A few words from the academy. *International Regional Science Review, 28*, 107–118.

Krygsman, S. & Dijst, M. 2001. Multimodal trips in the Netherlands: Microlevel individual attributes and residential context. *Transportation Research Record: Journal of the Transportation Research Board, 1753*, 11–19.

Larkin, B. 2013. The politics and poetics of infrastructure. *Annual Review of Anthropology, 42*, 327–343.

Latham, A. & Wood, P. R. H. 2015. Inhabiting infrastructure: Exploring the interactional spaces of urban cycling. *Environment and Planning A, 47*, 300–319.

Laurier, E., Brown, B. & McGregor, M. 2016. Mediated pedestrian mobility: Walking and the map app. *Mobilities: Traces of a Mobile Field: Ten Years of Mobilities Research, 11*, 117–134.

Law, R. 1999. Beyond 'women and transport': Towards new geographies of gender and daily mobility. *Progress in Human Geography*, *23*, 567–588.

Lee, S. & Talen, E. 2014. Measuring walkability: A note on auditing methods. *Journal of Urban Design*, *19*, 368–388.

Lefebvre, H. 2004. *Rhythmanalysis: Space, time, and everyday life*, London, Continuum.

Lopez, Russell P. & Hynes, H. P. 2006. Obesity, physical activity, and the urban environment: Public health research needs. *Environmental Health*, *5*, 1–10.

Lyons, G. 2004. Transport and society. *Transport Reviews*, *24*, 485–509.

Lyons, G. 2020. Walking as a service – Does it have legs? *Transportation Research. Part A. Policy and Practice*, *137*, 271–284.

Mateo-Babiano, I. 2016. Pedestrian's needs matter: Examining Manila's walking environment. *Transport Policy*, *45*, 107–115.

McFarlane, C. & Rutherford, J. 2008. Political infrastructures: Governing and experiencing the fabric of the city. *International Journal of Urban and Regional Research*, *32*, 363–374.

Merriman, P. 2019. Relational governance, distributed agency and the unfolding of movements, habits and environments: Parking practices and regulations in England. *Environment and Planning C: Politics and Space*, *37*, 1400–1417.

Metcalfe, R. & Dolan, P. 2012. Behavioural economics and its implications for transport. *Journal of Transport Geography*, *24*, 503–511.

Middleton, J. 2009. Stepping in time': Walking, time, and space in the city. *Environment and Planning A*, *41*, 1943–1961.

_____. 2010. Sense and the city: Exploring the embodied geographies of urban walking. *Social & Cultural Geography*, *11*, 575–596.

_____. 2011. "I'm on autopilot, I just follow the route": Exploring the habits, routines, and decision-making practices of everyday urban mobilities. *Environment and Planning A*, *43*, 2857–2877.

Mitullah, W., Opiyo, R. & Basil, P. 2019. *Transport and social exclusion in Kenya* [available at: https://intalinc.leeds.ac.uk/wp-content/uploads/sites/28/2019/06/Kenya-Scoping-study.pdf].

Moura, F., Cambra, P. & Gonçalves, A. B. 2017. Measuring walkability for distinct pedestrian groups with a participatory assessment method: A case study in Lisbon. *Landscape and Urban Planning*, *157*, 282–296.

Murray, L. 2020. The walking commute. In: Jensen, O. B., Lassen, C., Kaufmann, V., Freudendal-Pedersen, M. & Gøtzsche Lange, I. S. (eds.) *Handbook of urban mobilities*, Abingdon, Routledge, 109–117.

O'Dell, T. 2009. My soul for a seat: Commuting and the routines of mobility. In: Shove, E., Trentmann, F. & Wilk, R. R. (eds.) *Time, consumption and everyday life: Practice, materiality and culture*, Oxford, New York, Berg, 85–98.

Ogilvie, D. & Hamlet, N. 2005. Obesity: The elephant in the corner. *British Medical Journal*, *331*, 1545.

Parlette, V. & Cowen, D. 2011. Dead malls: Suburban activism, local spaces, global logistics (report). *International Journal of Urban and Regional Research*, *35*, 794.

Patterson, I., Pegg, S. & Omar, W. R. W. 2017. Walking to promote increased physical activity. In: Hall, C. M., Ram, Y. & Shoval, N. (eds.) *The Routledge international handbook of walking*, Abingdon, Routledge, 274–287.

Pikora, T., Giles-Corti, B., Bull, F., Jamrozik, K. & Donovan, R. 2003. Developing a framework for assessment of the environmental determinants of walking and cycling. *Social Science & Medicine*, *56*, 1693–1703.

Pivo, G. & Fisher, J. D. 2011. The walkability premium in commercial real estate investments. *Real Estate Economics, 39*, 185–222.

Potter, J. 1997. Discourse analysis as a way of analysing naturally occurring talk. *Qualitative Research: Theory, Method and Practice, 2*, 200–222.

Prillwitz, J. & Barr, S. 2009. Motivations and barriers to adopting sustainable travel behaviour [available at: https://ore.exeter.ac.uk/repository/handle/10036/104987].

Public Health Agency. 2020. Take the next step [available at: https://www.publichealth.hscni.net/sites/default/files/2020-02/Take%20the%20Next%20Step%20Booklet%2001_20.pdf].

Ram, Y. & Hall, C. M. 2017. Walkable paces for visitors: Assisting and designing for walkability. In: Hall, C., Ram, M. & Shoval, Y. (eds.) *The Routledge international handbook of walking*, Abingdon, Routledge, 311–329.

Ramsden, H. 2017. Walking & talking: Making strange encounters within the familiar. *Social & Cultural Geography: Negotiating Strange Encounters: Conceptualising Conversations Across Difference, 18*, 53–77.

Ravaisson, F. 2008. *Of habit, London*, New York, Continuum.

Saelens, B., Sallis, J. & Frank, L. 2003. Environmental correlates of walking and cycling: Findings from the transportation, urban design, and planning literatures. *Annals of Behavioral Medicine, 25*, 80–91.

Sallis, J. F., Owen, N. & Fisher, E. B. 2015. Ecological models of health behavior. In: Glanz, K., Rimer, B. K. & Viswanath, K. (eds.) *Health behavior and health education: Theory, research, and practice* (4th ed.), San Francisco, CA, Jossey-Bass, 465–486.

Sauter, D., Pharoah, T. M., Tight, M., Martinson, R. & Wedderburn, M. (2016). International walking data standard: Treatment of walking in travel surveys internationally standardized monitoring methods of walking and public space. Measuring walking [available at: https://files.designer.hoststar.ch/hoststar10546/file/1-international_walking_data_standard_version_aug_2016.pdf].

Schwanen, T. 2011. Car use and gender: The case of dual-earner families in Utrecht, the Netherlands. In: Lucas, K., Blumenberg, E. & Weinberger, R. (eds.) Bingley, Emerald Group Publishing Limited, 151–171

Schwanen, T. 2016. Geographies of transport I: Reinventing a field? *Progress in Human Geography, 40*, 126–137.

Schwanen, T., Banister, D. & Anable, J. 2012. Rethinking habits and their role in behaviour change: The case of low-carbon mobility. *Journal of Transport Geography, 24*, 522–532.

Schwanen, T. & Nixon, D. 2019. Urban infrastructures: Four tensions and their effects. In: Schwanen, T. & Kempen, R. V. (eds.) *Handbook of urban geography*, Northampton, MA, Edward Elgar Pub.

Sheller, M. & Urry, J. 2006. The new mobilities paradigm. *Environment and Planning A, 38*, 207–226.

Shove, E. 2010. Beyond the ABC: Climate change policy and theories of social change. *Environment and Planning A, 42*, 1273–1285.

Simmel, G. 1971. *On individuality and social forms: Selected writings*, Chicago, London, University of Chicago Press.

Smith, G., Sarasini, S., Karlsson, I. C. M., Mukhtar-Landgren, D. & Sochor, J. 2019. Governing mobility-as-a-service: Insights from Sweden and Finland. In: Finger, M. & Audouin, M. (eds.) *The governance of smart transportation systems: Towards new organizational structures for the development of shared, automated, electric and integrated mobility*, Cham, Springer, 169–188.

Sniehotta, F. F., Presseau, J. & Araújo-Soares, V. 2014. Time to retire the theory of planned behaviour. *Health Psychology Review*, *8*, 1–7.

Southworth, M. 2005. Designing the walkable city. *Journal of Urban Planning and Development*, *131*, 246–257.

Springgay, S. & Truman, S. E. 2019. Walking in/as publics: Editors introduction. *Journal of Public Pedagogies*, *4*, 1–12.

Star, S. & Ruhleder, K. 1996. Steps toward an ecology of infrastructure: Design and access for large information spaces. *Information Systems Research*, *7*, 111–134.

Star, S. L. 1999. The ethnography of infrastructure. *American Behavioral Scientist*, *43*, 377–391.

Steuteville, R. 2019. Walkability indexes are flawed. Let's find a better method [available at: https://www.cnu.org/publicsquare/2019/01/10/walkability-indexes-are-flawed-lets-find-better-method1].

Sunstein, C. R. 2019. *How change happens*, Cambridge, MA, MIT Press.

Swyngedouw, E. 2004. *Social power and the urbanization of water: Flows of power*, Oxford, Oxford University Press.

Thaler, R. H. & Sunstein, C. R. 2009. *Nudge: Improving decisions about health, wealth and happiness*, London, Penguin Books.

Transport for London. 2004. *Making London a walkable city: The walking plan for London*, London, Mayor of London.

_____. 2005. *Improving walkability*, London, Mayor of London.

_____. 2008. *Richmond and Twickenham set to become more legible for pedestrians* [available at: https://tfl.gov.uk/info-for/media/press-releases/2008/november/richmond-and-twickenham-set-to-become-more-legible-for-pedestrians].

_____. 2018. *Walking action plan: Making London the world's most walkable city*, London, Mayor of London.

_____. 2020. *The planning for walking toolkit*, London, Mayor of London.

Tranter, P. & Tolley, R. 2020. *Slow cities: Conquering our speed addiction for health and sustainability*, Oxford, Elsevier.

Ujang, N. & Muslim, Z. 2014. Walkability and attachment to tourism places in the city of Kuala Lumpur, Malaysia. *Athens Journal of Tourism*, *2*, 53–65.

Uteng, T. P. & Cresswell, T. 2008. *Gendered mobilities*, Aldershot, Ashgate.

Varela, F. 1992. The reenchantment of the concrete. In: Crary, J. & Kwinter, S. (eds.) *Incorporations*, New York, Zone.

Verplanken, B., Aarts, H. & Van Knippenberg, A. 1997. Habit, information acquisition, and the process of making travel mode choices. *European Journal of Social Psychology*, *27*, 539–560.

Warin, M., Moore, V., Davies, M. & Turner, K. 2008. Consuming bodies: Mall walking and the possibilities of consumption. *Health Sociology Review: Re-Imagining Preventive Health: Theoretical Perspectives*, *17*, 187–198.

Watson, A. & Drakeford-Allen, D. 2016. 'Tuning out' or 'tuning in'? Mobile music listening and intensified encounters with the city. *International Journal of Urban and Regional Research*, *40*, 1036–1043.

Yoshii, Y. 2016. Preserving alleyways to increase walkability of historical Japanese cities. *Procedia, Social and Behavioral Sciences*, *216*, 603–609.

3   Walking as social differentiation

In January 2013, Mitra Anderson spent two days with the global architect of urban public space, Jan Gehl, during his two-week international study tour of Melbourne. Anderson interviewed Gehl at various points during the tour about his perspectives on urban public space and what makes a liveable city. As Gehl reflected upon the transformation of Melbourne's public realm, he observed how in recent years it had gone from being 'a city where we once rushed to the office and back home again ("like ants to their various places and when they are finished they go down like ants down in the hole again"), to "a city which really is very inviting for promenading, and for lingering and sitting and enjoying, and looking at the girls or whatever you do."'[1] Gehl's comments prompt a series of questions and concerns in relation to this invitation to 'promenade' and/or 'linger'. Who is included and who is excluded from this vision of walkable spaces in cities? In what ways do different forms of walking exclude certain groups? How do different conceptions of walking work to produce particular forms of inclusionary and exclusionary politics?

In 2014, an edited video went viral of a young woman walking through the streets of New York City over a 10-hour period. The secretly filmed footage documents selected examples from over 108 instances of forms of sexual harassment by men towards the female actress. The film was distributed by Holloback!, an international movement concerned with anti-harassment measures in public spaces such as city streets. Following the wide circulation of the film, criticisms were made relating to how the editing of the footage promoted racial stereotypes in showing encounters with a disproportionate number of black men. These criticisms of the video portraying harassment along clear racialised lines are evidence of the pressing need to consider the intersectionality of social identities in any engagement with the politics of everyday walking practices. Yet, while remaining cognisant of these critiques, it is a powerful piece of film that clearly shows the ways in which women are frequently subjected to forms of harassment as they walk city streets in their everyday lives.[2] In Gehl's interview he claims, 'man was made to walk', with no mention of women until he proposes his vision of urban space where it is not only 'inviting' but also acceptable to 'look[ing] at the girls'. A gendered politics emerges from such claims in which women continue to be the object of the male pedestrian gaze through everyday urban walking practices.

This chapter examines the socially differentiated nature of everyday walking practices. In Chapter 1, I drew attention to bodies of work that emphasise the emancipatory nature of urban walking in relation to the democratic possibilities of public space (Jacobs, 1972; Sennett, 1970) yet here I argue that walking is as much about inequalities as it is to do with emancipation. More precisely, I contend that inequalities are not only produced through pedestrian practices, but that focusing on the 'simple act of walking' also brings many of the broader inequalities of contemporary urban life into sharp focus. As such, how people move on foot through urban space provides a pertinent empirical and conceptual entry point for examining the significance of the politics relating to how people appropriate urban space.

In what follows, I pay particular attention to the differential nature of walking in the city and associated inequalities. More specifically, I argue that there are inequalities associated with how walking is planned for and imagined through interventions such as traffic planning that privileges circulation through the timing and positioning of traffic signals or through more informal processes such as on-street parking where cars mount the pavement and obstruct pedestrian access (see Chapter 2 on pedestrian infrastructures). This is a particularly pressing concern, given that the significance of the exclusions and inequalities which manifest themselves through everyday walking practices have traditionally been overlooked by academics, policymakers, and practitioners. I argue that how people appropriate space in their everyday lives matters and it is important that we have the conceptual tools to understand the politics emerging from everyday walking practices. By politics, I mean how equality is enacted in specific times and places by different groups and people.

In this chapter I seek to counter the frequently disembodied and 'depoliticised' conceptions of walkability in transport and health arenas. This is not simply a lack of empirical data but a level of abstraction that makes it very difficult to see where an everyday subject fits into these understandings. In focusing on the complex inter-relations between class, race/ethnicity, gender, age, and disability I pay specific attention to a social politics of walking relating to the ways in which certain pedestrian types/forms/practices are deemed more important/significant than others. For example, the promotion of walking as a leisure activity in cities as opposed to being concerned with the everyday pedestrian mobilities of homelessness (see Hall and Smith, 2011; Jackson, 2015). In considering the politics of everyday urban walking it becomes of even more significance to further stress how walking and 'walkability' is not one thing to all people. To date, there remains a distinct lack of sustained attention to how walking intersects with social difference (Warren, 2017). As such, I focus here upon the intersectionality of social identities in relation to the everyday politics of urban walking.

I begin by examining the inequalities that emerge from how walking is imagined and planned for. It is not only 'utilitarian' forms of walking, such as the commute, that dominate urban and transport policy (see Chapter 2); walking for leisure is also privileged in many policy debates in relation to wider discourses of concerns with health (Berrigan et al., 2012; Green, 2009; Grénman and Räikkönen, 2017), well-being (McGinn et al., 2007; O'Donovan, 2015), and the rise in Euro-American contexts of adult and child obesity (Olabarria et al., 2014). I argue that these versions of urban walking are underpinned by privileged experiences that have become central to many of our understandings of everyday walking. I move on to return to Gehl's invitation to 'promenade' and/or 'linger' in urban public spaces and use the politics of the street as a lens through which to examine the inclusions and exclusions produced through everyday walking practices. The inequalities emerging from the differentially experienced nature of walking are significant, as they have implications for the types of walking positioned and promoted as being important within debates relating to how cities are planned for and understood. The chapter concludes by exploring walking as a form of protest in making the argument that urban walking not only produces inequalities through the ways it is imagined and practiced but it also makes visible broader inequalities in contemporary urban life.

Exclusionary walking imaginaries

I begin here from a starting contention, following Dawney (2011), that social imaginaries are produced through bodily practices and technologies and are both social and material. The ways in which walking is imagined and how these imaginaries are produced are of great significance to the exclusionary nature of pedestrian practices. The promotion of walking for leisure is central to many visions of creating walkable and pedestrian friendly cities, yet leisure walking practices more broadly have a long history of being exclusionary towards certain groups. For example, in 1932, over 500 walkers trespassed en-masse, walking from Hayfield to Kinder Scout, to protest for the 'right to roam'. The majority of walkers were working class people from Manchester who opposed the ways in which access to large parts of the UK countryside was experienced predominantly by the privileged classes. Although the 'white' landscape of the English countryside (Agyeman and Spooner, 1997) is now well recognised as discouraging a diversity of groups from recreational participation in activities such as walking for leisure, these forms of exclusion are still very much in evidence (see Chakraborti and Garland, 2014; Department for Environment, Food and Rural Affairs, 2019, Landscapes review).

Tourist practices across the globe are intrinsically linked to different ways of walking with the performance of pedestrian practices being central to how tourism spaces are constructed (Edensor, 2000; Reed, 2002; Solnit, 2014). For example, in Sarmento's (2017) study of tourists' walking rhythms in the medina of Tunis, he illustrates 'the complex manner in which tourist bodies, rhythms and urban forms intersect within the contemporary city that contributes to the constructed city itself' (295). He draws upon a combination of methods including mobile ethnography, or what he refers to as 'shadowing-at-a-distance', with accounts by travellers posted on internet travel forums. Sarmento argues that understanding the ways in which tourists move on foot through the alleys and streets of the medina is central to understanding the actual medina itself. In particular, he highlights how the performance of tourist walking practices strongly relates to Orientalist imaginaries (Said, 1978) of how the tourist industry positions Islamic–Arab cities as simultaneously exotic, magical, chaotic, dangerous, and violent. This is reflected not only in how tourists 'walk, move and engage with unknown terrain, but also the extent to which they look for or avoid interaction and encounter' (298).

In a related vein, Goh (2014) examines the politics of rhythm and memory surrounding urban walking in Singapore. He considers the ways in which the state has not only attempted to plan a 'walkable city' through urban redevelopment but through organised events and programmes to encourage specific ways of walking. Goh focuses on both the state promotion of heritage trails and the inaugural 2006 Singapore Biennale of international contemporary art. In doing so, he argues that the state appropriates historical space through these walking trails and downplays the 'collusion between the developmental state and global capital' (18). He turns to the work of the artist Amanda Heng to illustrate how her walking performances are a critical intervention in questioning the state's 'spatial production of the global city' (16) through the city's heritage trails. Both the medina in Tunis and the heritage trail in Singapore are examples of how the ways in which walking is imagined, in these cases in the context of leisure and tourism, produces particular versions of urban space. Such instances are significant as they illustrate how an exclusionary politics can emerge from how walking is imagined that has implications for who is, and who is not, welcome in such spaces and the encounters that may, or may not, unfold.

There is a long association between walking and other playful practices in the city that are also underpinned by privileged experiences, which in turn produce exclusionary imaginaries of urban walking. For example, the concept of the flâneur originated in the work of the French poet Charles Baudelaire (1964) and revolves around the concept of a gazing, male individual wandering through the public spaces of the city in a detached, ironic manner. In his writings, Walter Benjamin develops his own urban consciousness via the wanderings of the 19th-century flâneur: 'cities fascinated him [Benjamin] as a kind of organization that could only be perceived by wandering or by browsing' (Solnit, 2014: 197). Amin and Thrift (2002) point out how 'for some it is precisely the flâneur's sensibility linking space, language and subjectivity that is needed to read cities' (11). However, the flâneur is a privileged figure who is present, especially in the writings of

Benjamin, in the boulevards of European cities, making it possible to question the broader relevance of this approach for different climatic and cultural contexts.

Psycho-geography can also be understood as a means of engaging with, and often attempting to map, the ambiance and 'softer' dimensions of the city. There is a strong tradition of psycho-geography that has developed from the Situationist movement (see Debord, 1967; Pinder, 1996; Wollen, 1989) to more recent engagements in the work of Sinclair (2003, 2006) on his walks around different areas of London. The Situationists considered the 'dérive' as a key dimension in the construction of the psycho-geography of a city whereby a drifting motion around and through the city represented a political statement against rational, ordered, capitalist urban space. The marked growth of British psycho-geography is reflected in its visibility as a research tool, form of activism, and artistic practice (see Morris, 2019; Pinder, 2011; Richardson, 2015; Smith, 2015). For example, Bonnett (2017) is concerned with 'enchantment' in psycho-geographical walking and how magic is a site of potential creation and subversion for artists and writers. In exploring a rich array of British psycho-geographical forms and three London literary examples, he illustrates how 'magic is being used to conjure an aura of yearning, delight and transformation, as well as to offer practices of critique and disorientation that challenge the predictability of the ordinary landscape' (480). Bonnett highlights the contested nature of the term 'magic' while pointing to it as an active process that requires one to enact, perform, and conjure. He concludes by suggesting that identifying a 'longing for enchantment' across psycho-geographical walking 'calls forth a sense of possibility'. Yet, this raises questions as to who exactly this 'sense of possibility' emerges for. The voices which are included and excluded in this vision are a political question particularly as much psycho-geographical work traditionally lacks a diversity of voices.

The gendered nature of these more 'playful' aspects of walking is evident. Simonsen (2004) draws attention to the work of Buck-Morss (1989) on the writings of Benjamin and his arcades project, and Wolff (1985, 1994) on the invisibility of the female flâneuse, in highlighting 'a visualism and a gender-bias' in the figure of the flâneur. Simonsen continues to emphasise this 'visualism and a gender-bias' through describing how 'when the flâneur "goes botanising on the asphalt", he does so as a detached spectator and his visions are mediated through a male gaze, objectifying women as part of the urban landscape' (47). Bowlby (1991) describes flâneurie not only as male privilege in monetary terms but how women are also excluded 'a priori' due to them being an essential object of the male gaze (209).

Feminist critiques have also been levelled at psycho-geography in terms of the exclusionary nature of associated walking practices. For example, Heddon and Turner (2012) argue that the philosophical writings underpinning current understandings of walking as a performative practice are shaped by 'an implicitly masculinist ideology' which frames walking 'as individualist, heroic, epic and transgressive' (224). They propose that these underpinnings marginalise other forms of walking and through their writing seek to address the invisibility of women in much of this work. In a similar vein, Elkin (2017) also highlights the little attention that has been drawn to women writing about walking in the city. She argues that throughout history women have engaged with the city through many mediums including writing, photography, and film-making. She proposes that the concept of the flâneur should be re-considered in relation to the specificities of women's experiences as opposed to attempting to fit them into this masculine concept. This point is developed by Murali (2017) in their proposition of a manifesto to decolonise walking that recognises the dominance of walking narratives from European and American cities and the invisibility of 'minority bodies in the street' (85). They argue that decolonising walking is not simply providing an alternative race or gendered reading of urban pedestrian practices but a consciousness that examines 'the emergence and development' (86) of categories of race, gender, and sexuality in this context. Arora (2020) also emphasises the importance of shifting the centre of the dialogue around decolonising walking away from Europe and North America. In a related vein, Warren (2017) goes to great lengths to challenge the 'masculine, secular, and Euro-centric body' (802) that historically dominates understandings of everyday urban walking in her analysis of the everyday experiences of Muslim women migrants in urban public space. Warren (2017) pays specific attention to the walking interview as a method and is concerned with the lack of attention to the ethnic, gendered, and moral dimensions of its adoption. She argues that how '...geographical investigation into walking intersects with social difference has been surprisingly neglected' (787). She goes on to highlight the pressing need for more empirical engagements with 'the everyday socio-spatial practices of diverse, marginalised and/or vulnerable groups' (788).

In recent years, there has been a growing interest in mobility and aging, including concerns with how older age intersects with other differentiated socio-spatial processes (Schwanen and Páez, 2010). While walking has featured prominently in such work (see Bowering, 2019; Nijs and Daems, 2012; Schwanen et al., 2012), it is writings on children's independent mobility where the some of the most significant insights are located concerning the differential experiences of pedestrian practices in relation to life course. For example, Horton et al. (2014) explore the significance of 'just walking', where it is anything but simply walking, to children and young people's everyday mobilities. They challenge the taken-for-granted nature of how much walking is positioned in the work of social and cultural geographers and argue that it warrants much closer critical reflection. In similar terms to Clement and Waitt (2017) in their work on parenting mobilities, who acknowledge that more work needs to be done in understanding everyday walking beyond their focus on 'affluent white women' (1199), Horton et al. (2014) make calls for empirical and conceptual work on walking to more closely consider 'the constitution of diverse social and cultural inclusions and exclusions via walking practices' and how this intersects with the 'geographies of age, gender, class, ethnicity, disability, family or friendship' (112). In what follows, I take note of this call in examining more closely the inclusions and exclusions emerging from everyday pedestrian practices. I argue that the significance of walking in the city needs to be understood in these terms and not simply in relation to concerns with health and low-carbon mobilities.

Street life and pedestrian inequalities

At this point I want to return to Gehl's invitation to 'promenade' and/or 'linger' in urban public spaces and examine in more depth who is included and excluded from this vision of walkable spaces in cities. I probe these concerns further through considering the notion of sidewalks/pavements.[3] In addition to tracing the development of sidewalks/pavements, I draw upon the work of urban scholars on sidewalks/pavements, and the politics emerging from them being understood as spaces of circulation where only certain types of 'lingering' are tolerated. To 'stop and linger' is a differentially accessed privilege, which links to broader concerns with what types of walking are positioned and promoted as being important within debates relating to how cities are planned for and understood.

The sidewalk/pavement is a relatively new invention developed as a way of separating and managing different transport modes in the advent of the motor age. Bonham (2006) focuses on the Australian city of Adelaide to situate her discussion of the hierarchies which emerge from the 'separating out, classifying, and ordering travel practices in relation to their efficiency' (58). She explains how through the late 19th and early 20th century, efficient movement was the guiding principle for how street space was organised and urban traffic ordered. Separating pedestrians onto sidewalks was the result of such an ordering. For Blomley (2011), sidewalks have evolved as a powerful form of urban governance in relation to pedestrian flow. In his book *Rights of passage: Sidewalks and the regulation of public flow*, Blomley uses the term 'pedestrianism' as a means of denoting this logic, while tracing how the sidewalk can be understood in relation to the 'the promotion and facilitation of pedestrian flow and circulation, predicated on the appropriate arrangement of people and objects' (2). He argues that sidewalks have been neglected in the context of regulation, law, and order yet are central to how public space is conceptualised and regulated.

Sidewalks have long been focused on to situate wider concerns with how cities are imagined and planned for. For example, in her hugely influential book, *The death and life of the great American cities*, Jacobs (1972) stressed the importance of streets and sidewalks to the public life of cities. She emphasised the significance of the unplanned interactions of strangers, and the role these interactions play in maintaining safe urban areas. She argued that a well-used street is a safe street and how successful city neighbourhoods are those that use the presence of strangers as a safety measure in increasing the 'number of eyes' on the street. However, concerns with the ways in which sidewalks are governed and planned for lends itself as only a partial lens for understanding the everyday politics of walking and have less to offer in considering the micropolitics of everyday walking practices. This is more broadly symptomatic of a wider urban studies literature that is disembodied, lacks an in-depth engagement with the urban subject, and considers walking, without question, to be a positive urban practice. For example, in the work of French social theorist, Michel de Certeau, walking is framed as a form of urban emancipation that opens up a range of democratic possibilities. It is within this context that walking can be understood as a distinctly political act. De Certeau (1984) rejects the notion that pedestrians are shaped by urban space and control and explores walking as a mode of political resistance. He distinguishes between the 'strategies' of the powerful in their production of space and the 'tactics' of pedestrians in resisting these forms of control. In similar terms, Bridge (2004) emphasises how 'a walk can thread together diverse locations and situations and otherwise disrupt the proper geography of the city dictated by the rational plan' (123).

Other discussions of the democratic potential of cities also frame walking in relation to these emancipatory terms of reference. For example, Sennett (1970) draws upon the nature of encounters between strangers when walking and negotiating public space in the city. He considers the social heterogeneity of public urban spaces to offer opportunities of unpredictable but progressive social encounters, while Macauley (2000) also reflects upon the possibilities of pedestrian practices: 'urban and suburban walking raise issues related to trespassing and transgression, of crossing and marking the limit of the allowable. Walking de-limits boundaries, removes and reinstates a line or time with regard to permissibility or pass-ability' (207). The above examples are reflective of what Wilson (1992) refers to as 'the optimistic scenarios of Jacobs and Sennett' that 'date from a more optimistic time, the 1960s' (151). Yet this romanticism concerning the potential of walking in cities continues today across fields of academia, policy, and practice. Inattention to the embodied subject and the romanticisation of walking matters. For how people appropriate urban space in their everyday lives is an important concern and one for which we need an appropriate vocabulary for understanding and thinking about the associated politics of these walking practices. This is especially so as there is little evidence of how walking practices relate to how equality is enacted in specific times and places and the emergent/associated power relations. This is the social politics of everyday urban walking.

Considering how walking is socially and materially co-produced enables account to be taken of the significance of walking subjects. For example, Wilson (1992) highlights the tension, contradictions, and complexities surrounding women's experiences of walking in the city where she draws specific attention to how the city is a place of excitement and opportunity for women and not just a place to be feared. She describes how her mother planted within her 'a conviction of the fateful pleasures to be enjoyed and the enormous anxieties to be overcome in discovering the city' (1). She goes on to argue that 'women's experience of urban life is even more ambiguous than that of men, and safety is a crucial issue. Yet it is necessary also to emphasise the other side of city life and to insist on women's right to the carnival, intensity and even the risks of the city' (10). Bridge and Watson (2003) describe not only some of the fears for women negotiating urban public space by foot but also women who found freedom roaming the streets. They argue that 'what these contradictions reveal is the contradictory nature of urban public space for women' (370). Wilson (1992) points out how women have learnt to negotiate 'the contradictions of the city in their own particular way' (8). It can therefore also be argued that, like public space, the same can be said of the 'contradictory nature' of walking in the city.

However, concerns with sexual harassment and fear in the city are not the only ways in which the exclusionary gendered dimensions of walking emerge. As I highlight in Chapter 6, recent work on parenting mobilities, and more specifically mothering, also illustrates the gendered experiences of walking with young children. For example, Clement and Waitt's (2017) study of families journeying together builds upon the work of Boyer and Spinney (2016) by considering parenting mobilities with a focus on walking in the Australian city of Wollongong. The research brings the relationality of walking to the centre of the analysis through an assemblage reading of motherhood. In particular, they highlight the gendered nature of not only moments of care on foot but also moments of play in their participants everyday journeys. In my own research on new parents, care, and urban austerity (see Middleton and Samanani, 2021), the significance of walking is in evidence most significantly in relation to sleep. For as one mother noted in her diary:

> Much more relaxed to stay in Oxford rather than spending 15-20 mins in traffic trying to get out of the city with Holly screaming. It seems crazy to spend 24/7 trying to tend to her every need and keep her happy only to have to let her cry whilst driving. Very stressful and the low points this week all involved the car.
>
> High points are having my partner around to take Holly in the sling when we go for walks as during the week it feels like we are attached a lot of the time with feeding as sling walks.

For Lily, walking is central to her weekly routine as a new mother as she contrasts walking to the 'stressful' experiences of being in the car with her baby. In describing her and her partner's use of the sling she also makes available the significance of the mother-baby-assemblage to her walking practices. However, as Lily reflects upon the relief of her partner using the sling on walks at the weekends, what also becomes clear is the labour involved in caring for an infant and how this remains a very gendered form of care where women are disproportionately 'attached', both physically and emotionally, to their children. The labour becomes apparent through their everyday walking practices.

Yet, there remains an inadequacy in the conceptual tools emerging from these understandings as a means of engaging with the inequalities associated with everyday walking when we turn our focus away from the European contexts. For many cities, sidewalks/pavements do not exist, with different users competing for road space and pedestrians having to negotiate with multiple other forms of transport. This makes many aspects of pedestrian mobilities that are taken for granted in European countries, such as being able to cross the road safely, an everyday challenge. For example, Ogendi et al. (2013) highlight how pedestrians are over-represented in road traffic injuries and deaths in Nairobi. While examining community initiatives in London and Sao Paulo to improve walking and cycling infrastructures, Nixon and Schwanen (2018) draw attention to women from a low-income barrio in peripheral Sao Paulo who consider their neighbourhoods to be too dangerous for walking. Such experiences do not reflect the romanticised visions of walking for leisure and something to be enjoyed frequently promoted in European countries. For many, walking is the antithesis of what it means to be modern and, as such, walking is not a choice (although the above discussion of the gendered dimensions of walking already indicates that walking is not always a choice for women in most geographical contexts) and that we cannot presume that people's everyday walking experiences relate to forms of leisure or pleasure. This is no more apparent than when we take the work of street vendors. Such a focus brings starkly into view how urban inequality emerges in the lived experience of walking (see Martinez Rodriguez, 2019) where moving on foot for many is a last resort.

Informal street vending is prevalent across many non-Euro-American cities (Bandyopadhyay, 2016; Lin, 2018; Msoka, 2005). This work is frequently itinerant as a means of avoiding surveillance and policing. The significance of walking relates to restrictions frequently placed by city-level policies on stationary vending on streets and sidewalks. Such a position is underpinned by modernist visions of the city where streets and sidewalks are spaces of flow and circulation. This results in many vendors forced into sustaining an economic livelihood characterised by mobility in order to avoid penalties, such as a substantial fine or being arrested, imposed by city authorities. These restrictions are in large part a response to avoiding street vending taking up space and shaping pedestrian practices on the sidewalk. Eidse et al. (2016) explore such concerns through examining the everyday mobilities of street vendors in Hanoi, Vietnam, following a ban on street vending in 62 streets and 48 public spaces. They argue that street vending is an important but conceptually neglected concern, especially in socialist contexts. The authors set out to examine these everyday mobile experiences by drawing upon Cresswell's (2010) six facets of mobility (motive force, route, speed, rhythm, experience, and friction) and Kerkvliet's (2009) notion of everyday politics. The adoption of these approaches is to help understandings of 'the mobility politics affecting itinerant vendors compared to their stationary counterparts' (Eidse et al.: 344). They provide the reader with glimpses of the rich in-depth ethnographic work upon which the piece is based, contrasting sharply with the 'geographical bias' towards Euro-American contexts in work on the politics of the street. However, the theoretical promise of the piece in combining the work of Cresswell and Kerkvliet is less explicit and there remains a sense that there is so much more to be said about the politics emerging from the everyday experiences of these street vendors and their need, not desire, to keep becoming mobile.

The significance of walking to the everyday mobilities of informal street vending also emerges in the work of Esson et al. (2016) on mobile livelihood strategies in the Ghanaian capital city of Accra. In focusing on how urban residents use different modes of transport for income-generating activities, they highlight not only the creative adaptions people have to make in order to navigate a poor quality and often unreliable transport system but also the inequalities emerging from these practices. Through their empirical work they trace how many 'itinerant workers' create 'a sense of order and routine' (187) through the rhythms emerging from them developing their own timetables. For two of their informants, Kwesi and Evans, walking was central to their income-generating activities in avoiding law enforcement. For as Kwesi explained:

> I go to distant places because when you sell on the streets you will be arrested. I walk when I am selling. So when someone sees the goods and he likes it then he buys it'. While he would prefer to be stationary, with a fixed place to sell his goods, at present being sedentary is a risky strategy unless he can identify a space he perceives as safe. Evans' experience indicates the problem itinerant sellers can face when stopping: 'I laid my goods on the road side. The security men threw our things away yesterday. They are giving us a big problem. (187).

Esson et al. (2016) develop a holistic approach that brings transport, mobility, and livelihoods into dialogue with their empirical findings and in doing so question the appropriateness of conceptual frameworks emerging from the field of mobilities for understanding non-'Western' contexts. They contend that 'in this era of postcolonial and poststructural theorizing, it is essential to take due account of diverse conditions on the ground in different world regions' (187).

The work of Martínez Rodríguez (2019) is a step forward in taking account of such concerns in relation to how inequalities emerge from everyday pedestrian practices. In focusing on urban walking in Santiago de Chile, Martínez Rodríguez specifically examines how the lived experiences of walkers differ in relation to urban inequality (see also Jirón, 2010).

She makes a specific call for research on walking to take more account of a greater diversity of pedestrian practices. Through collecting accounts of walking through Santiago, the work of Martínez Rodríguez makes a valuable contribution to diversifying Anglophone engagements with urban walking while providing a new perspective on the lived experience of urban inequality in relation to everyday pedestrian practices.

In the next section I extend thinking on the exclusions emerging from everyday urban walking practices with the contention that focusing on pedestrian movement in the city also makes visible broader inequalities in contemporary urban life. More specifically I examine pedestrian protests to illustrate this point.

Urban inequalities and walking as protest

As a 12-year-old in the late 1980s I moved with my family from the UK to Southern California for a year due to my father's work. I began to attend the local junior high school and made the 20-minute journey from home to school on foot. I joined a small handful of students who walked to school, with the vast majority driven by their parents. There were few pavements/ sidewalks and much of the route involved walking along the sides of the road/on grass verges with cars driving past. The car-dominated culture of North America and poor walking environment has been well documented (see Dennis and Urry, 2009; Handy and Clifton, 2001). As I highlight in Chapter 2, the built environment does matter to our everyday experiences on foot and, relatedly, has an inevitable significant impact on the number of children who walk to school (Buehler and Hamre, 2015; Wen et al., 2008). However, for my 10-year-old sister, her reluctance to walk to school had little to do with the lack of pedestrian infrastructure but the link she had made between the photographs of missing children printed on milk cartons' she saw over breakfast and her fear of being abducted on the way to school. This small example further illustrates that our experiences of everyday walking relate to so much more than simply the built environment or 'pedestrian infrastructures' but the ways in which walking is materially *and* socially co-produced and imagined.

It was on my first few days walking to school that I learned of jaywalking. In Norton's (2007) historical account of jaywalking in the broader context of the dawn of the motor age in the United States, he reflects upon the etymology of the term: 'A jay was a country hayseed out of place in the city. By extension, a *jaywalker* was someone who did not know how to walk in the city' (342). Norton argues that legitimising jaywalking as a disreputable practice was key to US city streets becoming spaces of circulation for the motor car at the expense of pedestrians (see also Millington, 2014, on jaywalking). As I arrived opposite the school, there was a choice. You could walk down to the traffic lights to cross the road or save a few minutes by crossing during a gap in the traffic nearer to the school entrance. Although none of the children I walked with had ever been reprimanded by police for crossing illegally, I could not shake from my 12-year-old head the fear of being arrested each time we took the short cut across the road. Looking back, my fear, sense of wrong doing, and the severity of consequences I imagined possible were completely out of proportion to my actions. As a middle-class white kid, it was highly unlikely the police would ever intervene had they witnessed me crossing the road and committing a 'jaywalking' offence. However, in 2014 I was reminded of that journey to school and the politics and inequalities associated with the practice of jaywalking.

On 9 August 2014, an 18-year-old black man called Michael Brown was fatally shot by Darren Wilson, a 28-year-old white policeman in Ferguson, Missouri. It is alleged that the altercation that took place over Wilson's gun, which resulted in Brown's death, was initiated by Wilson instructing Brown and his friend to stop walking in the road and move onto the sidewalk. Although Wilson was cleared of Brown's murder, as investigators concluded there was no evidence to suggest that he wasn't in fear of his life, there is clear evidence that many initiations of jaywalking offences are racially motivated. For example, in Ferguson between 2011 and 2013, 95% of those charged with offences relating to 'manner of walking along roadway' were black (United States Department of Justice Civil Rights Division, 2015). While in 2019, 90% of jaywalking tickets in New York City were issued to black and Hispanic people (Kuntzman, 2020). In her account of the relationship between black mobilities and guns, Nicholson (2016) refers to this as a case of 'walking while black'. Following Brown's death, race riots and protests occurred in Ferguson and the broader Black Lives Matter movement gained momentum in protesting against US police violence against black people.

A political and socio-economic context of continued racial tensions and growing socio-economic inequalities has been one from which an increasing number of social movements, political activism, and civil demonstrations/ unrest have emerged. There is a long history of marching as a form of political expression with protests on foot having been adopted in multiple guises as a means of frequently providing a public stage for marginalised voices. For, as Bairner (2011) highlights, 'urban protest inevitably combines a desire to access the street with the use of that access for numerous other political objectives' (377). Yet considering pedestrian protests in more depth also provides illuminating insights into the inclusions and exclusions people experience in contemporary urban life.

On 14 June 2017, a fire tore through a residential tower block in the London borough of Kensington and Chelsea with devastating consequences. Seventy-two people lost their lives and 229 people escaped but lost their homes in the blaze. Combustible cladding on the exterior of the building contributed significantly to the speed the fire spread, while a single staircase fire escape was also attributable to the subsequent loss of life. Several months after this shocking tragedy, local residents initiated a silent walk on the 14th of every month to demonstrate against the injustice experienced by the community at the hands of the local authorities and central government. As I write, the Grenfell Silent Walk takes place each month and will continue to do so until the needs of those impacted by the disaster are met. From the silence surrounding each march emerges a powerful affective atmospheric force that visibly moves both those taking part and those bearing witness to this mass pedestrian act of local residents. For the organisers and participants of the march, walking in silence is a symbolic act whereby the silence of the march reflects the silence of the state in responding to not only the fire but also the widespread UK housing crisis, which, with increasing inequalities and a lack of affordable housing, the geographer Danny Dorling (2014) argues is the defining crisis of our times. The collective force of walking together is exemplified in this public display of solidarity.

In his writings on walking, memory, and protest, Back (2017) draws attention to the links between pedestrian practices and racial politics:

> In the turbulent and violent years of the last quarter of the twentieth century many marches took place along these south-east London streets. Walking here was much more than an everyday stroll; rather political struggles against racism were embodied in ways of moving through the city on foot. Offered here is a spatial story: one that creates anti-racist space by walking through a profoundly post-colonial place and feeling the traces of its history. (23)

Back's emphasis on the spatial stories of the anti-racist marches he traces through the streets of south-east London strongly resonates with the Grenfell Silent Walks. In the aftermath of the Grenfell fire, there was a dominant discourse in sections of the centre/right mainstream media that this was a human disaster and should not be politicised. This damaging rhetoric is still being shamefully peddled, evidenced in the comments of Conservative MP Jacob Rees Mogg, who claimed that residents who did not evacuate the building failed to use 'common sense'. Rees Mogg was asked on LBC radio to respond to the suggestion that 'in part the tragedy was caused by either racism or policies of class'. He replied, 'I don't think it's anything to do with race or class' (Merrick and Woodcock, 2019). Yet, the tragic events that unfolded at Grenfell were political not least due to the distinct racial and class politics at the heart of the deep-seated inequalities experienced by these local residents. For, as Moore (2017) wrote in the immediate weeks following the fire, 'Grenfell is as political as it gets. To deny that is an insult to the dead and an assault on the intelligence of the living'.

The routes of the Grenfell monthly silent walk tell a spatial story of marked racial and class inequalities. Following the six-month anniversary of the fire, the initial route around North Kensington, where the tower block was located, was altered to include wealthy areas south of the borough.

For the march organisers, it was considered an important political state-ment to take the walk through the prosperous streets of South Kensington. The Town Hall located in the south of the borough marked the beginning of the new routed march. Over time, the organisers hope to grow the march and extend its route through housing estates across London to further ignite the debate around the neglect of these buildings and how building regulation laws need to change. The Grenfell fire, and its devastating consequences, magnified the extreme class and racial inequalities that continue writ large across global cities such as London (cf. Brenner and Keil, 2006; Sassen, 2001), while the question remains of how people can live in the richest bor-ough in the UK but in unsafe housing. The Grenfell Silent Walks tell part of a tragic spatial story. The pedestrian acts of solidarity unfolding from the march serve to keep this story at the forefront of public consciousness until there is visible change in addressing the shocking neglect local communities have experienced at the hands of local and central government.

In her evocative cultural history of walking, Solnit (2014) paints a vivid picture of citizens coming together in the public spaces of cities as a form of public expression and political protest. For, as she highlights, 'Walking itself has not changed the world, but walking together has been a rite, tool and reinforcement of the civil society that can stand up to violence, to fear, and to repression. Indeed, it is hard to imagine a viable civil society with-out the free association and the knowledge of the terrain that comes with walking' (xii). There are multiple historical and contemporary examples of protest walks, too numerous to include here, ranging from the Civil rights marches led by Dr Martin Luther King in 1950s America to the recent pro-democracy demonstrations organised by students in Hong Kong. While the Grenfell Silent Walks speak to Solnit's concerns with resistance, there is more at stake as people gather together each month in solidarity. The calls of the protesters for justice also provide a sense of togetherness and comfort. These walks are about the intense emotions and affects produced through people consoling each other as they move together on foot. In the Antipode[4] plenary lecture at the RGS-IBG Annual conference in 2019, the sociolo-gist Les Back engaged with the conference themes of trouble and hope. He described in rich detail the Grenfell Silent Walks as an example of hope and connection. In his thought-provoking lecture, he reflected on the powerful affects produced through the silence of the walks as 'there is something allu-sive and impossible about cities without sound' (Back, 2020).

Solnit (2014) contrasts the expressive and political dimensions of 'parades, demonstration, protests, uprisings, and urban revolutions' to other 'practi-cal' everyday forms of walking (216)[5] and through this discussion of citizens on the street she raises three points that have particular resonance for con-sidering the everyday politics of urban walking. The first is the significance of walking to mobile forms of political protest in, and through, urban public spaces. Solnit highlights how '[such] walking is a bodily demonstration of political or cultural conviction and one of the most universally available forms of public expression' (217). The importance of walking to forms of political expression is also echoed in Reger's (2015) writing on SlutWalks (see also Kaygalak-Celebi et al., 2019, and Markwell and Waitt, 2009, on gay pride marches). The SlutWalk movement began as a protest to explanations of violence towards women being attributed to their appearance, with one particular Toronto Police officer suggesting that women 'should not dress like sluts' to avoid being sexually assaulted (Reger, 2015: 85). Many protesters on these walks deliberately wear clothing that some would deem as 'slut'-like attire, namely short skirts and tight tops. In Reger's critique of the use of the word 'slut' in relation to the emergence of SlutWalks, she draws explicit attention to the 'action and agency' of the practice of walking in these social movements and how 'when only the word "slut" resonates, the protest message is flattened, dropping out the sense of agency in reclaiming sexuality and fighting patriarchy' (96).

The second point of significance to draw here from Solnit's (2014) engagement with walking and forms of political expression is the attention she places on the pedestrian protesters themselves. In making a distinction between how a protest or procession is a 'participants' journey, while a parade is a performance with audience' (215), Solnit highlights the experiences of the walking subjects who take part in marches or demonstrations on foot. This is important, for when we ignore or overlook walking subjects we universalise pedestrian experiences, which in turn risks sidelining/losing the politics of pedestrian practices. Solnit's final point to emphasise are the material relations emerging from forms of political expression on foot. In developing through the pages of this book a conceptualisation of walking and walkability as being socially and materially co-produced, this concern is of particular relevance. Solnit makes explicit links between the material form of the city and how this relates to the ways in which people come together in urban spaces. It is clear the built environment matters, with Solnit contrasting riots in the urban sprawl of Los Angeles with 'old-fashioned' cities with boulevards that facilitated people congregating on the streets to parade, process, and demonstrate. In doing so, she argues how people write 'history with their feet' (223). However, a more implicit concern is also present in the attention she draws to material objects. Through a personal account of her witnessing the Day of the Dead procession through 24th Street in the Mission District of San Francisco, Solnit (2014) describes the 'huge crosses draped in tissue paper' and 'people carrying candles'. Material objects are thus central to this liminal festival celebrating life and death.

On 21 January 2017 over a million women descended on Washington DC, with estimates of between 3.3 and 4.6 million women (Waddell, 2017) taking part in similar protests across the globe, against the 45th newly elected US President and his anti-women statements and behaviour. The goal of the marches was to support legislation and policies in relation to human rights, women's rights, LGBTQ rights, worker's rights, healthcare and immigration reform, racial equality, and the environment. In the introduction to a special issue of *Gender, Place and Culture* focused on the significance of the Women's March, Moss and Maddrell (2017) highlight how the organisers of the march sought to be inclusive by bringing together diverse groups regardless of gender or gender identity. However, several contributions to the special issue challenged this inclusivity in examining the ways in which the solidarity of the march was also undermined. For example, Rose-Redwood and Rose-Redwood (2017) focus on the Women's March in Victoria, British Columbia, to explore questions of who is excluded and marginalised from these forms of protest, particularly in relation to the dominance of whiteness.

In a similar vein, Boothroyd et al. (2017) argue that the Women's March on Washington 'reproduced hegemonic notions of female bodies' (717) through its use of the pussyhat[6] and its role in reducing gender identity to female genitalia. The conceptual lens of engaging with walking as a socially and materially co-produced practice is of particular value here. For the women who took part in the march, despite broader representations in the media, their experiences were not universal. This further re-enforces one of the key arguments I present through this book: not all walking is the same. Furthermore, this approach makes visible the significance of the use of pussyhats by some demonstrators in terms of the affective forces and politics produced through the social and material assemblages emerging from the march. Moss and Maddrell (2017) highlight how expressions on the Women's March through 'material and discursive icons and memes, support and contest the meaning of inclusion for solidarity' (616). Understanding these expressions includes taking seriously the significance of material objects to how the politics of walking as a form of protest unfolds. More broadly, the emergent politics of pedestrian protests can be rendered more visible through engaging with how walking practices, in these contexts, are both socially and materially co-produced. This serves more generally to illustrate broader urban inequalities.

Conclusion

Our experiences of everyday mobility differ and are intimately connected to social identities. It is through focusing on differentiated experiences of mobility that the significance of the politics of mobility become apparent. For example, Cresswell (2010) ascertains the importance of considering the politics of specific aspects of mobility. He proposes 'breaking mobility down into six of its constituent parts (motive force, velocity, rhythm, route, experience, and friction) in order to fine-tune our accounts of the politics of mobility' (17). However, the experiences of different people and the ways they walk through the city are a neglected concern across a multitude of engagements with urban walking. For example, the differential nature of walking the city and its associated inequalities receive little attention in urban planning contexts focused on active travel or

low-carbon mobilities. Yet, as this chapter has illustrated, not all walking can be reduced to being one thing for all people. The overlooking
of differentiated experiences on foot is important, for there is a distinct
everyday politics emerging from how people appropriate space in their
urban walking practices which requires attention. Considering walking
as an intersectional practice in relation to the complex inter-relations
of class, race/ethnicity, age, disability, and gender makes visible the
inequalities emerging from the different types, forms, and practices of
walking. Urban walking is as much about the inclusions and exclusions
relating to how it is planned for and imagined and everyday experiences
on foot as it is about the frequency of pedestrian trips, the perceived
'walkability' of a particular built environment, and the health benefits
of walking.

Reflecting on gendered experiences of walking, the racialised spatial stories of walking protests, and the literal need (not choice) to keep
walking as an itinerant vendor on the city streets of Hanoi and Accra are
just a handful of glimpses into which we can see how urban inequalities
are both embedded within, and emerge from, different urban pedestrian
practices. Yet, even within these walking practices, experiences of moving on foot cannot be universalised. Accounts of the Women's March
illustrate just this point. Furthermore, in this chapter I have emphasised
the need to diversify work on walking beyond research and planning
discourses situated in the 'West'. This is a point that Schwanen (2018)
makes in the context of transport and mobilities research more broadly,
when he invites us to seriously consider the 'worlding' of geographical research on transport and mobility. He points to the work of Jazeel
(2014) in the importance of 'unlearning the kind of conceptual language
[and methodological practice] that we already know as social scientists'
(98). Schwanen argues that this does not mean rejecting outright the
dominant 'Western' approaches to engaging and understand transport
and mobility but that we should critically engage with and challenge the
'myriad effects of their circulation and invocation' (469).

These are important considerations and I will turn to the implications of
this invitation for how walking is understood in the context of contemporary urban life in the concluding chapter of the book. However, at this point
I want to continue my exploration into the socially differentiated nature of
everyday walking practices. In this chapter I have considered inclusions and
exclusions emerging from how everyday walking is imagined and practised
along with the wider urban inequalities it makes visible in relation to well-
established debates on social identities such as race and gender. I now
turn my attention to subtler forms of social differentiation, hierarchy,
and privilege through a discussion of in-depth empirical data from my
research where the significance of everyday walking practices, beyond
well-rehearsed concerns with low-carbon and healthy cities, emerges in a
range of different contexts.

Notes

1. See http://assemblepapers.com.au/2013/06/13/cities-for-people-jan-gehl/.
2. In 2016, Nottinghamshire Police become the first UK police force to recognise misogyny as a hate crime. This includes catcalling and wolf-whistling at women in the street (see http://www.nottinghamwomenscentre.com/wp-content/uploads/2018/07/Misogyny-Hate-Crime-Evaluation-Report-June-2018.pdf). The Everyday Sexism Project is a UK-based initiative that aims to catalogue women's daily experiences of sexism and includes multiple examples of on-street harassment (see https://everydaysexism.com).
3. I use the terms sidewalks and pavements interchangeably depending on the geographical context I am describing. For example, in the UK, the word 'pavement' is used when referring to the paved path that runs alongside the street to provide pedestrian access to the front of buildings. In other geographical contexts, such as North America, the term 'sidewalk' is adopted. It is important to note that in many countries, such as Brazil and Kenya, sidewalks/pavements are an uncommon urban design feature.
4. *Antipode: A Radical Journal of Geography* is a peer-reviewed journal 'that push[es] Geography's critical edge, intending to engender the development of a new and better society' (see https://onlinelibrary.wiley.com/page/journal/14678330/homepage/productinformation.html).
5. It is worth noting that not all forms of walking together are about resistance to power. For example, military parades throughout history are demonstrative of order, structure, and compliance. Although it could be suggested that the French military services marching band's choreographed performance of a Daft Punk medley in front of President Trump and President Macron on Bastille Day in 2017 was a micro-expression of resistance as the bemused President Trump displayed his utter disdain throughout!
6. The Pussyhat Project was founded by Jayna Zweimand and Krista Suh in the run-up to the Women's March. The intention was to encourage people to knit a pink hat in the form of cat ears as a response to a television recording of Trump claiming he liked to 'grab them [women] by the pussy' (BBC, 2016). He subsequently attributed this revolting statement to 'locker-room banter'.

References

Agyeman, J. & Spooner, R. 1997. Ethnicity and the rural environment. In: Cloke, P. J. & Little, J. (eds.) *Contested countryside cultures: Otherness, marginalisation and rurality*, London/New York, Routledge, 197–217.

Amin, A. & Thrift, N. J. 2002. *Cities: Reimagining the urban*, Cambridge, Polity.

Arora, S. 2020. Walk in India and South Africa: Notes towards a decolonial and transnational feminist politics. *South African Theatre Journal*, 1–20.

Back, L. 2017. Marchers and steppers: Memory, city life and walking. In: *Walking through social research*, London, Routledge, 21–37.

_____. 2020. Hope's work. *Antipode* [available at: https://onlinelibrary.wiley.com/page/journal/14678330/homepage/productinformation.html].

Bairner, A. 2011. Urban walking and the pedagogies of the street. *Sport, Education and Society: New Directions, New Questions? Social Theory, Education and Embodiment*, *16*, 371–384.

Bandyopadhyay, R. 2016. Institutionalizing informality: The hawkers' question in post-colonial Calcutta. *Modern Asian Studies*, *50*, 675–717.

Baudelaire, C. (translated and edited by Jonathan Mayne). 1964. *The painter of modern life, and other essays*, Oxford, Phaidon.

BBC. 2016. *US election: Donald Trump sorry for obscence remarks on women.* 8th October [available at: https://www.bbc.co.uk/news/election-us-2016-37594918].

Berrigan, D., Carroll, D., Fulton, J., Galuska, D., Brown, D., Dorn, J., Armour, B. & Paul, P. 2012. Vital signs: Walking among adults – United States, 2005 and 2010. *Morbidity and Mortality Weekly Report*, *61*, 595.

Blomley, N. K. 2011. *Rights of passage: Sidewalks and the regulation of public flow*, Abingdon, Routledge.

Bonham, J. 2006. Transport: Disciplining the body that travels. *The Sociological Review*, *54*, 57–74.

Bonnett, A. 2017. The enchanted path: Magic and modernism in psychogeographical walking. *Transactions of the Institute of British Geographers*, *42*, 472–484.

Boothroyd, S., Bowen, R., Cattermole, A., Chang-Swanson, K., Daltrop, H., Dwyer, S., Gunn, A., Kramer, B., McCartan, D. M., Nagra, J., Samimi, S. & Yoon-Potkins, Q. 2017. (Re)producing feminine bodies: Emergent spaces through contestation in the Women's March on Washington. *Gender, Place & Culture: Emergent and Divergent Spaces in the Women's March: The Challenges of Intersectionality and Inclusion*, *24*, 711–721.

Bowering, T. 2019. Ageing, mobility and the city: Objects, infrastructures and practices in everyday assemblages of civic spaces in East London. *Journal of Population Ageing*, *12*, 151–177.

Bowlby, R. 1991. Walking, women, and writing: Virginia Woolf as flâneuse. *Tropismes*, *5*, 207–232.

Boyer, K. & Spinney, J. 2016. Motherhood, mobility and materiality: Material entanglements, journey-making and the process of 'becoming mother'. *Environment and Planning D: Society and Space*, *34*, 1113–1131.

Brenner, N. & Keil, R. 2006. *The global cities reader*, London, Routledge.

Bridge, G. 2004. Everyday rationality and the emancipatory city. In: Lees, L. (ed.) *The emancipatory city?: Paradoxes and possibilities*, London, Sage, 123–138.

Bridge, G. & Watson, S. 2003. *A companion to the city*, Malden, MA, Blackwell Publishing.

Buck-Morss, S. 1989. *Dialectics of seeing: Walter Benjamin and the Arcades Project*, Cambridge MA, London, MIT Press.

Buehler, R. & Hamre, A. 2015. The multimodal majority? Driving, walking, cycling, and public transportation use among American adults. *Planning – Policy – Research – Practice*, *42*, 1081–1101.

Chakraborti, N. & Garland, J. (eds.). 2014. *Rural racism*, Abingdon, Routledge.

Clement, S. & Waitt, G. 2017. Walking, mothering and care: A sensory ethnography of journeying on-foot with children in Wollongong, *Australia. Gender, Place & Culture*, *24*, 1185–1203.

Cresswell, T. 2010. Towards a politics of mobility. *Environment and Planning D: Society and Space*, *28*, 17–31.

Dawney, L. 2011. Social imaginaries and therapeutic self-work: The ethics of the embodied imagination. *Sociological Review*, *59*, 535–552.

De Certeau, M. 1984. *The practice of everyday life*, Berkeley, London, University of California Press.

Debord, G. 1967. *La société du spectacle*, Paris, Buchet/Chastel.

Department for Environment, Food and Rural Affairs. 2019. Landscapes Review [available at: https://www.gov.uk/government/publications/designated-landscapes-national-parks-and-aonbs-2018-review].

Dennis, K. & Urry, J. 2009. *After the car*, Cambridge, Polity.

Dorling, D. 2014. *All that is solid: How the great housing disaster defines our times, and what we can do about it*, London, Penguin Books.

Edensor, T. 2000. Walking in the British countryside: Reflexivity, embodied practices and ways to escape. *Body & Society*, 6, 81–106.

Eidse, N., Turner, S. & Oswin, N. 2016. Contesting street spaces in a socialist city: Itinerant vending-scapes and the everyday politics of mobility in Hanoi, Vietnam. *Annals of the American Association of Geographers: Geographies of Mobility*, 106, 340–349.

Elkin, L. 2017. *Flâneuse: Women walk the city in Paris, New York, Tokyo, Venice and London*, London, Vintage Books.

Esson, J., Gough, K. V., Simon, D., Amankwaa, E. F., Ninot, O. & Yankson, P. W. K. 2016. Livelihoods in motion: Linking transport, mobility and income-generating activities. *Journal of Transport Geography*, 55, 182–188.

Goh, D. P. S. 2014. Walking the global city: The politics of rhythm and memory in Singapore. *Space and Culture*, 17, 16–28.

Green, J. 2009. 'Walk this way': Public health and the social organization of walking. *Social Theory & Health*, 7, 20–38.

Grénman, M. & Räikkönen, J. 2017. Taking the first step: From physical inactivity towards a healthier lifestyle through leisure walking. In: Hall, C. M., Ram, Y. & Shoval, N. (eds.) *The Routledge international handbook of walking*, Abingdon, Routledge, 288–299.

Hall, T. & Smith, R. 2011. Walking, welfare and the good city. *Anthropology in Action*, 18, 33.

Handy, S. & Clifton, K. 2001. Local shopping as a strategy for reducing automobile travel. *Planning – Policy – Research – Practice*, 28, 317–346.

Heddon, D. & Turner, C. 2012. Walking women: Shifting the tales and scales of mobility. *Contemporary Theatre Review: Site-specificity and Mobility*, 22, 224–236.

Horton, J., Christensen, P., Kraftl, P. & Hadfield-Hill, S. 2014. 'Walking ... just walking': How children and young people's everyday pedestrian practices matter. *Social & Cultural Geography*, 15, 94–115.

Jackson, E. 2015. *Young homeless people and urban space: Fixed in mobility*, New York, Routledge.

Jacobs, J. 1972. *The death and life of great American cities*, Harmondsworth, Penguin.

Jazeel, T. 2014. Subaltern geographies: Geographical knowledge and postcolonial strategy. *Singapore Journal of Tropical Geography*, 35, 88.

Jirón, P. 2010. Mobile borders in urban daily mobility practices in Santiago de Chile. *International Political Sociology*, 4, 66–79.

Kaygalak-Celebi, S., Kaya, S., Ozeren, E. & Gunlu-Kucukaltan, E. 2019. Pride festivals as a space of self-expression: Tourism, body and place. *Journal of Organizational Change Management*, 33, 545–566.

Kerkvliet, B. J. 2009. Everyday politics in peasant societies (and ours). *The Journal of Peasant Studies*: Critical Perspectives in Agrarian Change and Peasant Studies, 36, 227–243.

Kuntzman, G. 2020. 'Jaywalking While Black': final 2019 numbers show race-based NYPD crackdown continues. *Streetsblog NYC.* 27th January [available at: https://nyc.streetsblog.org/2020/01/27/jaywalking-while-black-final-2019-numbers-show-race-based-nypd-crackdown-continues/].

Lin, S. L. 2018. We work like ants… we avoid being troublemaker. *International Journal of Sociology and Social Policy*, *38*, 1024–1040.

Macauley, D. 2000. Walking the city: An essay on peripatetic practices and politics. *Capitalism Nature Socialism*, *11*, 3–43.

Markwell, K. & Waitt, G. 2009. Festivals, space and sexuality: Gay pride in Australia. *Tourism Geographies: Tourism Development and Perceptions in Australia and Canada*, *11*, 143–168.

Martínez Rodríguez, S. A. 2019. May I walk with you? Exploring urban inequality in everyday walking practices in Santiago de Chile. Unpublished PhD thesis, University College London [available at: https://discovery.ucl.ac.uk/id/eprint/10069934/].

McGinn, A. P., Evenson, K. R., Herring, A. H. & Huston, S. L. 2007. The relationship between leisure, walking, and transportation activity with the natural environment. *Health & Place*, *13*, 588–602.

Merrick, R. & Woodcock, A. 2019. Jacob Rees-Mogg sparks fury by blaming Grenfell victims for not leaving building. *Independent.* 5th November [available at: https://www.independent.co.uk/news/uk/politics/jacob-rees-mogg-grenfell-tower-fire-common-sense-kensington-a9185501.html].

Middleton, J. & Samanani, F. 2021. Accounting for care within human geography. *Transactions of the Institute of British Geographers*. 46, 29–43.

Millington, G. 2014. Jaywalking. In: Atkinson, R. (ed.) *Shades of deviance: A primer on crime, deviance and social harm*, Abingdon, Routledge, 17–20.

Moore, S. 2017. Grenfell is political. The right can't make that fact go away [available at: https://www.theguardian.com/commentisfree/2017/jun/26/grenfell-political-right-tory-john-mcdonnell-fire].

Morris, B. 2019. *Walking networks: The development of an artistic medium*, London, Rowman and Littlefield International.

Moss, P. & Maddrell, A. 2017. Emergent and divergent spaces in the Women's March: The challenges of intersectionality and inclusion. *Gender, Place & Culture*, *24*, 613–620.

Msoka, C. 2005. Informal markets and urban development: A study of street vending in Dar es Salaam, Tanzania. Unpublished PhD thesis, University of Minnesota.

Murali, S. 2017. A manifesto to decolonise walking: Approximate steps. *Performance Research: On Proximity*, *22*, 85–88.

Nicholson, J. A. 2016. Don't shoot! Black mobilities in American gunscapes. *Mobilities: Mobilities Intersections*, *11*, 553–563.

Nijs, G. & Daems, A. 2012. And what if the tangible were not, and vice versa? On boundary works in everyday mobility experience of people moving into old age: For Daisy (1909–2011). *Space and Culture*, *15*, 186–197.

Nixon, D. & Schwanen, T. 2018. Emergent and integrated justice: Lessons from community initiatives to improve infrastructures for walking and cycling. In: Butz, D. & Cook, N. (eds.) *Mobilities, mobility justice and social justice*, London, Routledge, 129–141.

Norton, P. D. 2007. Street rivals: Jaywalking and the invention of the motor age street. *Technology and Culture*, *48*, 331–359.

O'Donovan, H. 2015. *Mindful walking: Walk your way to mental and physical well-being*, Dublin, Hachette Books Ireland.

Ogendi, J., Odero, W., Mitullah, W. & Khayesi, M. 2013. Pattern of pedestrian injuries in the city of Nairobi: Implications for urban safety planning. *Bulletin of the New York Academy of Medicine*, *90*, 849–856.

Olabarria, M., Pérez, K., Santamariña-Rubio, E. & Novoa, A. M. 2014. Daily mobility patterns of an urban population and their relationship to overweight and obesity. *Transport Policy*, *32*, 165–171.

Pinder, D. 1996. Subverting cartography: The Situationists and maps of the city. *Environment and Planning A*, *28*, 405–427.

_____. 2011. Errant paths: The poetics and politics of walking. *Environment and Planning D: Society and Space*, *29*, 672–692.

Reed, A. 2002. City of details: Interpreting the personality of London. *Journal of the Royal Anthropological Institute*, *8*, 127–141.

Reger, J. 2015. The story of a slut walk: Sexuality, race, and generational divisions in contemporary feminist activism. *Journal of Contemporary Ethnology*, *44*, 84–112.

Richardson, T. 2015. *Walking inside out: Contemporary British psychogeography*, London, New York, Rowman & Littlefield International.

Rose-Redwood, C. & Rose-Redwood, R. 2017. It definitely felt very white': Race, gender, and the performative politics of assembly at the Women's March in Victoria, British Columbia. *Gender, Place & Culture: Emergent and Divergent Spaces in the Women's March: The Challenges of Intersectionality and Inclusion*, *24*, 645–654.

Said, E. W. 1978. *Orientalism: Western conceptions of the orient*, Harmondsworth, Penguin.

Sarmento, J. 2017. Tourists' walking rhythms: 'Doing' the Tunis Medina, Tunisia. *Social & Cultural Geography*, *18*, 295–314.

Sassen, S. 2001. *The global city*, Princeton, Princeton University Press.

Schwanen, T. 2018. Geographies of transport III: New spatialities of knowledge production? *Progress in Human Geography*, *42*, 463–472.

Schwanen, T., Hardill, I. & Lucas, S. 2012. Spatialities of ageing: The co-construction and co-evolution of old age and space. *Geoforum*, *43*, 1291–1295.

Schwanen, T. & Páez, A. 2010. The mobility of older people – An introduction. *Journal of Transport Geography*, *18*, 591–595.

Sennett, R. 1970. *The uses of disorder: Personal identity & city life*, New York, Knopf.

Simonsen, K. 2004. Spatiality, temporality and the construction of the city. In: Bærenholdt, J. O. & Simonsen, K. (eds.) *Space odysseys: Spatiality and social relations in the 21st century*, Aldershot, Ashgate, 43–62.

Sinclair, I. 2003. *London orbital: A walk around the M25*, London, Penguin Books.

_____. 2006. *Edge of the Orison*, London, Penguin Books.

Smith, P. 2015. *Walking's New movement*, Axminster, Triarchy Press.

Solnit, R. 2014. *Wanderlust: A history of walking*, London, Verso.

United States Department of Justice Civil Rights Division. 2015. Investigation of the Ferguson Police Department, 4th March [available at: https://www.justice.gov/sites/default/files/opa/press-releases/attachments/2015/03/04/ferguson_police_department_report.pdf].

Waddell, K. 2017. The exhausting work of tallying America's largest protest [available at: https://www.theatlantic.com/technology/archive/2017/01/womens-march-protest-count/514166/].

Warren, S. 2017. Pluralising the walking interview: Researching (im)mobilities with Muslim women. *Social & Cultural Geography*, *18*, 786–807.

Wen, L. M., Fry, D., Rissel, C., Dirkis, H., Balafas, A. & Merom, D. 2008. Factors associated with children being driven to school: Implications for walk to school programs. *Health Education Research*, *23*, 325–334.

Wilson, E. 1992. The invisible flâneur. *New Left Review*, *191*, 90–110.

Wolff, J. 1985. The invisible flâneuse: Women and the literature of modernity. *Theory, Culture & Society*, *2*, 37–46.

_____. 1994. The artist and the Flaneur: Rodin, Rilke and Gwen John in Paris. In: Tester, K. (ed.) *The flâneur*, London, Routledge, 111–137.

Wollen, P. 1989. The Situationist International. *New Left Review*, *174*, 67–95.

4 Walking rhythms and the politics of time

> Dust [a local bar] appears opposite the next junction, and not a moment too soon as this collection of small stages can be deceiving. It's not uncommon to feel one step on from where I actually am and, psychologically, that lengthens my journey.
>
> (Diary – Paul, Canonbury resident, Islington)

Why is 'this collection of small stages' deceiving to Paul; how does Paul 'psychologically' lengthen his journey; and why does he often 'feel one step on' from where he actually is? This chapter explores walking and its relationship with and through time and space. In particular, I engage with the temporal rhythms of walking while considering the conflicts and vulnerabilities that emerge from these everyday walking rhythms. In encouraging people to adopt active and low-carbon forms of transport, walking is often promoted within transport, health, and urban policy as a quick and efficient mode of transport. The significance of time and its influence on decisions to walk has been well documented in previous research (see Banister, 2011; Carlson et al., 2015; Joshi and Senior, 1998). Yet, across many conventional transport perspectives, walking is framed in relation to travel time being considered 'dead time' that people seek to minimise.

This chapter builds upon work that has sought to challenge dominant linear conceptualisations of time in transport planning and policy as nothing more than clock time passing. In line with research on 'productive travel time use' (Jain, 2011; Lyons and Urry, 2005; Singleton, 2019; Tang et al., 2018; Watts and Urry, 2008), I highlight how walking opens up the possibility of doing other things in a way that other forms of urban transport are unable to do. I further draw upon the in-depth research on everyday walking in London introduced in Chapter 2 to engage with experiential time, or people's sense of time, as a means of exploring the multiple temporalities that emerge out, and shape, people's everyday experiences on foot. In doing so, I use the writings of Adam (1990) and Bergson (1911/1998) to examine experiences of time and suggest that people become aware of their own duration when they are made to wait. The chapter moves on to argue that notions of rhythm provide a productive means of engaging with the multiple, mutually dependent, temporalities, and spatialities (see Massey, 2005) that emerge out of and shape people's everyday walking experiences. However, while there are other studies that engage with the spatio-temporal rhythms of walking (see for example Matos Wunderlich, 2008) and walking practices as a form of political resistance (de Certeau, 1984), far less attention has been paid to the conflicts and vulnerabilities of pedestrian rhythms in the context of concerns with active travel and low-carbon mobilities. As such, I argue that considering the rhythms of the walking body is essential for further understanding the differentiated nature of pedestrian experiences and associated inequalities of everyday urban walking.

Travel in-between time

Historically, there has been an emphasis on speed and efficiency within the wider transport arena where 'any opportunity to reduce journey-times is constantly sought, be it through increasing speed or by traveling on the fastest, or shortest, available route to a destination' (Harris et al., 2004: 6). Some would argue that the value of speed has been elevated to 'inalienable right, where the ability to travel as fast as possible is no longer questioned' (ibid). Furthermore, in transport appraisals, any savings in travel time represent a significant benefit of a scheme (Lyons and Urry, 2005) with time being considered a significant cost (Goodman, 2001; Tranter and Tolley, 2020). Since the mid-2000s there has been a growing recognition in both academic and policy arenas that travel time is not 'dead time' that people seek to minimise (see Sheller and Urry, 2006). This shift corresponds with thinking emerging from the mobility turn that challenges conventional perspectives of transport being a derived demand with no positive utility in itself 'by exploring how travel time can be, and is, being used "productively" as activity time' (Lyons and Urry, 2005: 257).

However, this interest in travel time is on passengers being transported by some other mode of transport such as a train or aeroplane, with specific attention on passengers' interactions with other passengers, wireless technologies, the view from the window, and objects in their bags (see for example Bissell, 2009; Watts and Urry, 2008, on trains; Laurier et al., 2008, on cars; Clayton et al., 2017, on bus travel; and Jain, 2011, on coach travel). Little attention has been paid to walking in the context of travel time use, particularly relating to working time. This reflects a wider tension around how walking time should be addressed. For example, in literature engaging with the value of travel time, walking (and cycling) is frequently considered as being part of calculations of total travel time as opposed to being dealt with as separate entities. While in much transport analysis of travel time, despite, for example, accounting for over 27% of all trips made in the United Kingdom (Department for Transport, 2018), walking is omitted altogether. Working-while-walking is not something typically associated with pedestrian movement (although see Chapter 3 and discussion of itinerant vendors), perhaps due to the everyday walker, unlike the train or aeroplane passenger, not being stationary.

Furthermore, there still remains in some academic and policy spheres an emphasis on speed and efficiency. For example, public health literature has drawn positive associations between walking speeds and durations and health outcomes (Pae and Akar, 2020; Saevereid et al., 2014; Sahlqvist et al., 2012), while TfL (2020) promote walking as 'surprisingly… quicker than expected'. Their website features maps detailing the number of steps and time it takes to walk between each tube station as a means of encouraging people to walk more frequently due to it being 'quicker' than taking the tube. Relatedly, in Melbourne, the city's Walking Plan (2014–2017) makes claims as to how a rich walking network 'reduces walking times' (City of Melbourne, 2014: 12). The 'time saving' attributes of walking and efficiency of certain routes are illustrative of how time is frequently considered a limited resource and currency not to be wasted (Bell, 1998; Thrift, 1977). However, in what follows I challenge the temporal assumptions of people's desire to move more quickly on foot through critically evaluating the premise that 'faster is seen to be better, as it achieves more in a given time frame' (Harris et al., 2004: 6). As such, the data discussed next highlights the significance of travel time for journeys on foot. For example, the following extract from Paul's diary illustrates how for many residents walking is a significant part of their day-to-day working life; they work while they walk:

> The next few minutes are lost to me. I must have walked past the theatre but didn't notice as my brain has switched to thinking about brochure spreads and advertisements. It's amazing how I suddenly start conceiving ideas about work whilst walking along – even without realizing it.
> Since I get paid to think, I wonder if I should be paid for my walking. But then again, I also get paid to be in by 9!!
> (diary – Paul, Barnsbury resident, Islington)

Clearly time is at issue for Paul and walking is taken to blur the boundary between walking time and working time. However, Paul's claim concerning a loss of time ('the next few minutes are lost to me') is accounted for in an interesting way. He demonstrates how he enters into time. He marks this in terms of what he 'must have' missed – 'the theatre'. What it is to be consciously aware is also used as a resource to demonstrate the noteworthiness of what he is claiming to be significant about this particular walk to work. These are constituted as more than random thoughts in the time it takes to walk to work. 'Brochure spreads and advertisements' are used to build the sense of experiencing time in the journey. These 'ideas' are presented as isomorphic, as exactly the equivalent to what it is to work. Indeed, as Paul claims, he works 'even without realising it', prompting him to raise the question of 'since I get paid to think, I wonder if I should be paid for my walking?'

Paul's account is orientated to issues also addressed in Lyons and Urry's (2005) study of travel time and their contention that 'commuting is not considered to take place in work time' yet 'implicitly it may already be that employers and employees are blurring the accepted boundary between the commute and time spent at work' (273) (see also Bissell, 2018 on commuting). However, Paul's account is also an example of how participants use time as a resource for reflecting upon other issues in the context of walking. In this instance Paul draws upon temporal concerns ('The next few minutes are lost to me') to make points about his identity as a hard-working, creative ('I suddenly start conceiving ideas about work'), media professional ('brochure spreads and advertisements'). In the interview extract below, connections are also made between productivity at work and the walk to work:

KRIS: Actually when I was working in London in the late 80s, there was a fundraiser at the English National Opera and she was very successful, very well-known and people still talk about her now even though she's left, one of the things everybody knew about her was that she walked to work, it took her 2 hours every morning and I was always intrigued by, you know the connection between that and the success she had fundraising. I think it's part of her success.

JM: And why do think that is, do you think that's in terms of the time she had to think?

KRIS: Yes, because I think when you take public transport it's, it's, you're kind of consumed by, well certainly I have in the past just before I started walking, consumed by, sometimes three 38 buses would go by and they won't stop at all so there's this anxiety of 'oh I'm getting late, I'm getting late', you don't have that when you walk. And seeing the tube you know, at Highbury and Islington, it gets so jammed packed and sometimes I have to wait for three, and I, when I lived with my sister in, in Hackney, I took the North London line and that's incredibly packed, I saw a couple of fights break out. So yes, so it's, I find, I found commuting here very tense (interview – Kris, Canonbury resident, Islington).

These comments were made in the context of Kris making a comparison with travelling on public transport to explain how walking expands time for him to think about work. He illustrates this further by attributing the success of a woman within his industry to her two-hour walk to work, something he is keen to emulate in his pedestrian activities. In her writings on the history of walking, Solnit (2014) draws attention to the relationship between walking and thinking time: 'walking allows us to be in our bodies and in the world without being made busy by them. It leaves us free to think without being wholly lost in our thoughts' (5). However, the significance of Kris's account not only lies with these observations concerning 'walking and thinking' or with how he positions the reliability of walking in contrast to the No. 38 bus, but the data also draws attention to the multiple rhythms and rhythmicity of walking. Kris illustrates how he is concerned by the time criticality of his journey and the anxiety this engenders in terms of full buses not stopping. Missed buses disrupt the 'regulated rhythms' of journeying in a timetabled manner to work. This disrupted temporal rhythm and what it is to be late ('oh I'm getting late, I'm getting late') is used to set up the contrast of commuting by public transport with walking to work ('you don't have that when you walk'). The all-consuming stress that Kris has encountered on public transport exemplifies an urban experience of temporality of regulated rhythms, yet the way in which he compares this experience with walking actually disrupts this common representation of urban dwelling as relentless, busy, and time compressed. I will return to how notions of rhythm assist in our understandings of everyday urban walking later in this chapter but first I wish to further consider how travel time is conceptualised.

Although there is a growing acknowledgement across academic and policy arenas (see Wardman and Lyons, 2016) that travel time is not 'dead time' but something that people use productively, travel time itself is still widely considered as a Cartesian form of linear temporality waiting to be filled with some form of activity (Bissell, 2007). Perhaps the most well-known manifestation of this understanding of movement and time was the emergence of time-geography in the 1960s in the work of Torsten Hägerstrand (1967, 1975, 1977). Hägerstrand and his colleagues' concern was the movement of individuals through time and space. Their understandings of human spatial behaviour were based on the premise that time and space are resources that enable individuals to carry out personal activities or projects. These movements were plotted on time-space graphs in their belief that mapping individual and institutional projects in time-space built up a contextualised understanding of how societies maintain and reproduce themselves (Latham, 2003). However, time-geography has received much criticism over the years, particularly by feminist and cultural geographers. For example, in work on women's role as carers, Davies (2001) argues that time-geography wrongly assumes time is a resource everyone has equal access to. In considering the 'rhythms of the city', Crang (2001) draws attention to, and makes specific comparisons between, the work of Hägerstrand and Lefebvre. Lefebvre's (2004) interest in rhythm stemmed from his project of rhythmanalysis and his concerns with the everyday:

> The everyday is simultaneously the site of, the theatre for, and what is at stake in a conflict between great indestructible rhythms and the processes imposed by the socio-economic organization of production, consumption, circulation and habitat (73).

Crang (2001) introduces 'the overlain multiplicity of [urban] rhythms' (189) with Lefebvre's notion of rhythmanalysis and uses this to critique the linear understandings of time and the everyday that frame the work of time-geographers such as Hägerstrand. In drawing upon Lefebvre to critique linear understandings of time, Crang's concern is 'the differences between lived and represented times' (187) as he explores the notion of 'experiential time-space' with the aim of offering 'a less stable version of the everyday, and through this a sense of practice as an activity creating time-space not time-space as some matrix within which activity occurs' (187). It is to these 'experiential time-spaces' that I now turn.

Experiential time and the politics of waiting

Adam (1995) highlights how clock time is deeply implicated in temporal understandings within contemporary 'Western' social sciences and how 'machine time has been reified to a point where we have lost touch with other rhythms and with the multiple times of our existence' (27). Historically, much of transport studies, and the transport policy it has informed, has been underpinned by understandings of time 'as a one-dimensional, uncomplicated, mechanistic and measurable concept' (Goodman, 2001: 50). Glennie and Thrift's (2009) historical account of timekeeping challenges such perspectives. Their approach argues against a singular history of clock time while stressing the importance of understanding the diversity of clock times. Levine (2003) draws attention to the decreasing focus across the social sciences on linear understandings of time, and how different forms of time emerge in the construction of phenomena. Rethinking the role of temporal concerns reflects what Adam (1990) refers to as a shift from 'events in time' to 'time in events'. As Levine (2003) explains:

> Time in events' prioritizes the search for the multiple times in any object of inquiry ... Of course, many of the times to be identified will have linear properties, but the intention is to focus on the multiplicity of times – rather than to concentrate on following the logic of one kind of time. (60).

For example, in Hagman's (2006) study of traffic jams and parking problems in Sweden he argues that the difference between waiting in traffic jams and searching for a car parking space has to do with how people experience time loss: 'in the queue, time slips away slowly. In searching for a parking spot ... we can count the seconds. Losing countable seconds is perhaps more annoying than losing uncountable hours' (72). Hagman's empirical focus has wider significance relating to the variability associated with how people experience time.

Yet more recently, Bastian (2017) has raised concerns as to how clock time has been positioned in such engagements and argues that 'across a wide range of cultural forms, including philosophy, cultural theory, literature and art, the figure of the clock has drawn suspicion, censure and outright hostility' (41). In contrast she suggests that what is meant by clock time should be opened up, that clocks are not apolitical, and how we need to pay closer attention to 'the complex ways clocks and clocktime are constructed' (3). For example, time-keeping 'etiquette' can vary across the globe, with being early considered as rude as being late in some countries, such as Japan. While notions of 'island time', associated with South Pacific Islands, relate to racist ideologies embedded within concerns with modernity and efficiency. The politics of clock time are incredibly important, yet in what follows I examine how pedestrians experience time, the significance of waiting, and why these temporal experiences matter to our understandings of walking in the city. Adam's (1990) notion of 'time in events' is drawn upon to make sense of the multiple temporalities associated with, which cut across, and emerge out of pedestrian movement.

The extract from this Barnsbury resident's diary describes a journey to work:

Monday 9th May
  Left house at 8.20 am and arrived at work at 9:00am.
  Again we managed to have a family outing to the coffee shop before I walked to work.
  I had to get to work for a conference call at 9:00 so I was in a bit of a rush to get to the office.
  I walked part of the way with my husband, he on his bike. Then had to speed up to get to the office on time. When I am in a rush I take the back road that runs parallel to Farringdon Road. I am not sure if it is quicker but it feels it.
  (diary – Gemma, Barnsbury resident, Islington)

Like many of the participants, Gemma's diary extract is an example of how issues associated with time are drawn upon as a resource by participants as they account for the multiple things and relations emerging out of and shaping their pedestrian movements. Issues concerning identity are of particular significance in terms of how people temporally frame distinctions they make about who they are in relation to others. For example, Gemma's reference to the 'family outing to the coffee shop' is done in relation to how they have made time to do it. This statement suggests that this 'family time' is something valued and done on a regular basis ('Again we managed'). The conference call that Gemma is 'in a bit of a rush to get to' is a further example of how temporal concerns frame other issues. In this instance it makes relevant Gemma's professional relationship to others. These workplace relationships are made significant in terms of the rush she is in on that morning. This is more than ascribing to Gemma some 'professional' occupation and her role being one of significant importance to participate in this call. The temporalities in her account are what makes available the multiplicities of Gemma's identity not just as a pedestrian who tempers the rhythm and pattern of her walk to work in relation to the situations she has to deal with, but also in terms of her accountabilities to others as a partner, family member, and office member.

As requested in her diary instructions, Gemma notes down the date, time, how long, and where she walked. These details reflect a linear representation of time within Gemma's walking patterns on that day. However, with Adam (1995) highlighting that 'the time of clocks and calendars is but one aspect of the many times that bear on our lives simultaneously' (16) and that 'the existence of clock time, no matter how dominant, does not obliterate the rich sources of local, idiosyncratic and context-dependent time awareness which are rooted in the social and organic rhythms of everyday life' (21), what other 'time' is 'in' this 'event' of Gemma walking to work? In addition to clock time, what can be described as 'collective' times (see Crow and Heath, 2002) also shape and emerge out of Gemma's pedestrian movements. The family outing to the coffee shop and the conference call at 9.00 reflect Adam's (1995) explanation of how 'a moment of "my" time is never just that. It is inseparable from "our" times, the times of the environment and the social collectivity' (19). What Gemma's account illustrates is how 'collective' and 'clock' times are inextricably linked to other times, times that Adam (ibid) describes as 'when' times which relate to cultural expressions of time and 'timings': 'Whilst the existence of clock time facilitates context independence and global standardisation, decisions about the timing of even the most habitual actions are made on a unique basis and with reference to a particular context' (22). These habitual 'timings' are reflected upon and made visible by Gemma as she was in 'a bit of a rush' and had to 'speed up to get to the office on time'.

'Collective times' are significant features of Durkheimian sociology. For Durkheim (1915/1976) all time is 'social time', and both time and space are collective representations. He equated time to 'an endless chart, where all duration is spread out before the mind, and upon which all possible events can be located in relation to fixed and determinate guidelines' (10), whereby 'society' itself imposed a temporal ordering on events or collective activities such as feasts and public ceremonies. However, as Gemma describes the route she takes when she is in a rush and how it is not necessarily quicker but because 'it feels it', her account bears little relation to the time-space budget she details, or 'an endless chart': she is not quantifying and measuring time, instead she 'feels it'. Thus, a further temporality becomes evident, that of 'experiential time' (see Gell, 1992, for critical review of various philosophical engagements of 'experiential time'). But how is it possible to understand and make sense of this 'feeling' of time Gemma describes? In other words, people's experiences of time passing?

Below, I turn again to Paul's diary entry (also discussed in Chapter 2), as it illustrates how it is through 'time in' the 'event' of waiting that can assist in understanding the experiential dimensions of time or how people become aware of time when they are made to wait. This exemplifies how multiple temporalities intersect and their relationship with, and through, pedestrian movement:

> This time last week I had just come back from holiday and my traffic light awareness function was not quite operating at its optimal capacity. You get a sense of timing when doing the same route each day. Where to cross and lose least time. Which corners are blind and make the heart race. And which order the traffic lights change in so you're one pace ahead of the infrequent walkers.
>
> Last week my timing was all over the place – now I'm back, a fact brought to my attention as Toby runs to follow me across the road. Point to note: I never run across the roads. It's just not dignified.
>
> (diary – Paul, Canonbury resident, Islington)

This extract from Paul's diary describes ways in which he negotiates the challenges of walking in his local area. Paul frames time as valuable in relation to his sense of 'where to cross and lose least time' but does more than describe a walking route characterised by speed and efficiency with time positioned as a scarce resource. The multiplicity of time and how different temporalities intersect is prevalent throughout his account. 'Clock times', 'collective times', and 'timings' mutually interact, both shaping and emerging out of his peripatetic movements. Furthermore, Paul draws upon temporal issues and concerns in positioning himself as a regular ('you're one pace ahead of the infrequent walkers') and accomplished ('I never run across the roads. It's just not dignified') pedestrian. However, more can be said about this 'sense of timing' that Paul describes. For example, how does he come to achieve it, how does it shape his temporal experiences, and what enables him to be become aware of it? Furthermore, what is it for Paul to be achieving 'optimal' walking and of what significance are the traffic light sequences, the road crossings, blind corners, infrequent walkers, and Toby's (Paul's friend) 'undignified' dash across the road to his 'sense of timing'? Paul's account could be read through the lens of a largely neglected notion of time called Kairos. With Ancient Greek origins, Kairos describes an opportune moment for action. Yet, it is by turning to French philosopher Henri Bergson that it is possible to begin to engage with how time is experienced and this 'sense of time' that Paul is referring to.

Bergson's concern was with time and in particular the experience of time passing. He used the notion of 'duration' to emphasise the continuity, irreversibility, and openness of time (Massey, 2005). In 'Time and Free Will', Bergson (1913/2001) argued conceiving of time as lived experience; 'duration' 'involves a succession of conscious states that enables the conscious being to "break" their absorption in the present moment' (Middleton and Brown, 2005: 62–63). Bergson (1946/1992) continued his work on 'duration' and used the notion of 'hesitation' or 'elaboration' to explain how people become aware of duration; how they experience time. He argued how:

> ...the living being essentially has duration; it has duration precisely because it is continuously elaborating what is new and because there is no elaboration without searching, no searching without groping. Time is this very hesitation, or it is nothing. (172)

It is the intersection of people and things, when people are made to wait, which reveals in this case, Pauls' 'sense of timing' during his walk to work. Bergson's use of 'hesitation' can be drawn upon to understand how Paul's return from holiday, his 'traffic light awareness function', the pace of infrequent walkers, and his friend Toby's dash across the road, intersect to illustrate the durational qualities of his journey and how time expands and contracts. What it is to 'wait' can be illustrated further by Bergson's (1911/1998) frequently cited example of mixing sugar and water:

> If I want to mix a glass of sugar and water, I must, willy nilly, wait until the sugar melts. This little fact is big with meaning. For here the time I have to wait is not that mathematical time ... It coincides with my impatience ... It is no longer something thought, it is something lived. (9–10)

Bergson's example highlights the relational durations of the sugar, water, and the observer and the unfolding experience of being made to wait. Adam (1990) also draws attention to time in relation to waiting: 'our awareness of the complexity of social time beyond the symbol and the quantitative measure can be enhanced by focusing on the phenomenon of waiting' (121). Paul's account demonstrates how 'knowing the times proper to things, processes, and events is, of course, fundamental to human time-reckoning systems' and how 'we have to know those times before we can recognise them as regular and consistent in relation to other that are faster and slower' (ibid: 122). It is thus waiting that is key to this knowledge. Yet, although Adam's makes reference to phenomenon of waiting, it is not developed at any length, particularly in relation to the actual experience of waiting (see Jeffrey, 2010, on the politics of waiting).

The remaining extract from Paul's diary entry further highlights how the 'event' of waiting reveals the lived and experiential dimensions of time:

> Toby doesn't usually walk this way and marvels at the length of the walk knowing I do it twice daily but we're not yet at the fire station so all's well.
>
> The fire station is a seminal landmark on the route home. Almost meaningless on the way to work, on the way back it is the point against which I can judge my trip. If I am beginning to flag before I get there,

then the last twenty minutes will be agony. Sail past it, however, and I know I'm full of beans.

(diary – Paul, Canonbury resident, Islington)

As Paul describes his journey home, he does not adopt a linear understanding of time to 'measure' his progress. Instead, this 'hesitation' as the fire station and his physicality intersect reveals his experience of time. The fire station has durational properties in terms of Paul's walking. This is special to him and not Toby. His awareness of the duration of his walk to work expands and contracts in terms of his sense of well-being at the fire station; in his words 'a seminal landmark'. But the fire station is more than a landmark as it intrudes into his rhythm of walking differently on the outward and homeward journeys. In his journey home Paul draws upon the fire station in more ways than solely revealing the durational properties of his movements on foot. He also uses it to judge how he will feel on the final stretch home; whether he will be in pain or cruising home with ease. As Bissell (2007) explores the 'corporeal experience of the event of waiting during the process of journeying' (277), he not only proposes that in order to do so 'a non-linear apprehension of qualitative temporality' (284) must be taken seriously, but also that moments of waiting heighten a self-awareness of the physicality of the body and how they 'may be considered to be intensely corporeal events: an awareness of one's own body in space' (286). The event of waiting at the fire station, however momentarily, reveals both Paul's sense of time, his physicality, and how they intersect.

Notwithstanding how Bergson's conceptualisations of time in terms of duration have assisted in my analysis of the relationship between walking and time, there are those who are critical of Bergson and others for privileging time over space (Latour, 2005; Massey, 2005). By his own admission in *Matter and memory* (1908/1991), Bergson emphasises the dualistic nature of his approach to time and space with respect to how pure time should not be confused with pure space. Massey (2005) also raises concerns in relation to this dualism:

> None of these philosophers has the reconceptualisation of space as their objective. More often, and in the context of wider debates, temporality is a more pressing concern. Over and again space is conceptualised as (or, rather, assumed to be) simply the negative opposite of time. (17)

However, Massey (2005) moves on to highlight how her issue is not with this 'prioritisation of time over space' but with 'the way we imagine space' (18). Massey is critical of the ways space is associated with representation in the writings of Bergson. She argues that conceptualising space in this way rendered it static, 'deprived of any dynamism, and radically counterposed to time' (21). Massey draws attention to how space, in like terms as to how time has been discussed in this chapter, can be conceived of as multiple, as always in process, and in relation to how it expands and contracts. However, Massey's key point is that time and space cannot be understood in isolation. In other words, her argument 'concerns the mutual necessity of space and time' and how 'it is on both of them together, that rests the liveliness of the world' (56). How then can pedestrian movement be understood in light of these conceptual concerns relating to time and space?

Engaging with concerns of duration and people's experiences of time on foot matter to understandings of walking, walkability, and the walkable city. As I have discussed, walking is frequently positioned across multiple arenas as accessible to all, quick, and efficient yet the variability of the temporalities emerging from everyday walking practices further emphasises how all walking is not the same. This seemingly obvious point is often overlooked. To ignore the ways in which everyday urban walking is socially differentiated results in the inequalities emerging from such practices also being neglected. This becomes particularly apparent when the potential conflicts and vulnerabilities associated with the walking body are examined. In the next section I extend this analysis of urban walking in order to address such concerns. In particular, I explore how the rhythms of walking bodies, in terms of the inter-relatedness of time and space, and how these can be disrupted, provide a lens to further illustrate the socially differentiated nature of everyday urban walking.

The rhythmicity of walking

> I'm thinking about my case tomorrow and what I'm going to say – this is easier to do when I'm walking for some reason. If you are cycling or running, you have to concentrate a bit more.
> (diary – Lucy, Canonbury resident, Islington)

In similar terms to how Paul and Kris describe their walk to work in the previous section, Lucy also explains how she utilises her daily pedestrian commute to think about work ('I'm thinking about my case tomorrow') and how she finds it 'easier to do' on foot than using other modes of transport. However, it is by turning our attention to notions of rhythm that further sense can be made of her account. Amin and Thrift (2002) point out how 'the rhythms of the city are the coordinates through which inhabitants and visitors frame and order the urban experience' and how 'the city is often known and negotiated through these rhythms and their accompanying ordering devices (traffic rules, telephone conventions, opening times, noise control codes)' (17). The significance of city rhythms to contemporary urban life has been explored in a variety of settings too numerous to examine in any detail here (see for example Smith and Hall, 2012, on the rhythms of urban outreach work; Lyon, 2016, on Billingsgate Fish Market; and Latham and Layton, 2019, on urban sociality and public space).

Lucy's account of 'walking and working' draws attention to the rhythms that characterise this particular dimension of her urban pedestrian experiences. For, as Highmore (2005) points out, 'rhythm in the form of pace, is a crucial ingredient to any text or any experience of the city, no matter how fast or slow that pace is' (141). It is the variations in 'pace' between cycling, running, and walking and what these rhythms afford in term of 'thinking time' that can be interpreted as illustrative in Lucy's diary extract. However, exploring the connection between walking, working, and rhythm still leaves open the issues of just how does one become aware of these rhythms; in other words, how are urban rhythms apprehended in the context of pedestrian movement? Mels (2004) highlights this very issue: 'while for obvious reasons rhythm is essential to musicians, poets, and dancers, how could rhythms be of any interest to human geographers? And if so ... how would we grasp those rhythms?' (4).

Lefebvre (2004) aimed to develop an awareness and appreciation of rhythm through the project of rhythmanalysis and his concerns with the everyday. Key to Lefebvre's understandings of rhythm was the importance of thinking space, time, and place as inter-related. He proposed that 'everywhere where there is interaction between a place, a time and an expenditure of energy, there is rhythm' (15). For despite rhythms being 'tentatively defined as movements and differences in repetition, as the interweaving of concrete times', they also imply 'a relationship of time to space and place' (Simonsen, 2004: 45). This conceptual shift was a necessary dimension of rhythmanalysis as it allowed Lefebvre to move from a dualistic to a dialectical analysis and enable 'notions and aspects to be likened to it that analysis too often keeps separate; times and spaces, the public and the private, the state-political and the intimate' (Grönland, 1998). Various critiques have been levelled at Lefebvre's project of rhythmanalysis including there being no clear method for its execution as a tool of analysis (Amin and Thrift, 2002) and it being an incomplete project framed by limited concepts and hypotheses (Simonsen, 2004). Lyon's (2018) introductory book to rhythmanalysis provides a more balanced position. Through a series of empirical examples, she explores both the limitations and possibilities of rhythmanalysis as a 'strategy of enquiry' (5) for research on the everyday.

Despite these shortcomings, I will now examine the potential of rhythmanalysis for assisting our understandings of exploring the relationship between space and time in the context of urban walking. For as Highmore (2005) points out: 'rhythmanalysis is an attitude, an orientation, a proclivity: it is not "analytic" in any positivistic or scientific sense of the term. It falls on the side of impressionism and description, rather than systematic data collecting' (150). Furthermore, Lefebvre (2004) stressed how rhythm is only an object for analysis in so far as the dynamic interdependencies of place, time, and energy are made visible in some way and how, in similar terms to Bergson (1911/1998) and his notion of what it is to wait in relation to experience of time, 'we are only conscious of most rhythms when we begin to suffer from some irregularity' (Lefebvre, 2004: 77). Thus, there are clearly ways in which Lefebvre's conceptions of rhythm can be drawn upon to further engage with resident's accounts of their everyday urban walking practices.

Like Paul, Kris, and Lucy, the following diary extract also draws attention to how walking enables people to think:

> I often have a walk if I need to think about something and walk with partner at the weekend – for a couple of hours.
>
>   Walking produces a metronome effect. Moving fast sometimes means that it is difficult to think something through. Moving slowly allows my mood to be more considered + therefore makes it easier to think.
>   (diary – George, London Fields resident, Hackney)

George's account makes a specific link between the speed he walks and how this relates to his ability to 'think something through'. What this illustrates is how people attend to the rhythms of their walking practices but not as an object in and of themselves. Rather, the mode of analysis is the way irregularities and differences in terms of the interdependencies of spatialities, temporalities, and corporalities are both discernable and become available as a resource in the production of accounts. In line with Bergson (1911/1998), it is the slowing down, or 'hesitation', that reveals the rhythmic and durational dimensions of George's movements on foot. George's account illustrates the multiple paces and rhythms that emerge out of, and shape, experiences on foot. Thus, rhythm is a way of understanding the multiple temporalities, spatialities, and corporalities of walking together. In other words, where there is rhythm of sorts, there is something to be said about time and space.

Elden (2004) also emphasises how rhythms are not the object of analysis themselves as he points to the role and significance that Lefebvre places on the body: 'the rhythmanalyst does not simply analyse the body as a subject, but uses the body as the first point of analysis, the tool for subsequent investigations. The body serves us as a metronome. This stress on the mode of analysis is what is meant by a rhythmanalysis rather than an analysis of rhythms' (xii). It is interesting to note how this understanding of rhythm is an entanglement of both the quantitative and qualitative and therefore transcends such a distinction. It is the concept of the body serving as, in George's own words, a 'metronome' that is reflected in his account. The significance of the body in the context of the rhythmicity of urban walking continues as this De Beauvoir Town resident reflects upon his experiences of walking at night:

JM: What's the longest you'd walk day to day as opposed to leisure walks?
DAVE: I'd do 5 miles, I can do more than 5 miles. I sometimes, John who works here with me, like we'll go to some drinks thing or something and we'll have a few drinks and then we'll walk back you know. And

we'll walk back from Soho or somewhere back here. It will be nighttime and it's just lovely. I love walking through London at night.

JM: You do? What do you like about it?

DAVE: I just like, I like being able to talk very loud and just the kind of stimulation of where you pass.

JM: How would you say that's different than at night?

DAVE: I just think at night you're not, there's no, you're allowed to just dawdle and piss about in a way you can't in the daytime. People don't, it's just that I don't [dawdle], because in the daytime I'm always doing something whereas at night I'm relaxed (interview – Dave, De Beauvoir Town resident, Hackney).

Dave responds to the question of the 'longest' he would walk day-to-day by making a claim in relation to a measurable distance of '5 miles'. In fact, Dave moves on to state how he 'can do more than 5 miles'. However, this distance is qualified in an interesting way. In accounting for the distance of his day-to-day movements on foot, Dave draws distinctions between walking during the day and at night in the city ('I love walking through London at night'; 'you're allowed to just dawdle and piss about in a way you can't in the daytime'; 'in the daytime I'm always doing something whereas at night I'm relaxed'). How Dave makes these distinctions can be understood in light of the relationship of night to day in the context of urban pedestrian experiences (see Chapter 5 on walking at night and Beaumont, 2015, for a detailed historical account of nightwalking). Notions of rhythm are made available by Dave as he draws upon them as a resource for accounting for how he can walk 'more than 5 miles' at 'nighttime'. For example, rhythmicity emerges as he makes a comparison with how you can 'dawdle' at night and how he is more 'relaxed' than during the day. Notions of rhythm are also drawn upon in making visible how space is sensually apprehended in relation to how Dave likes being able to 'talk very loud' and the 'stimulation of where you pass'.

Furthermore, Dave's observations can be understood in relation to the emphasis Lefebvre places on the body. For example, as Dave describes how he likes to walk home following numerous drinks after work due to the 'stimulation', being able to 'talk very loud', how he can 'dawdle', and is 'relaxed', this reflects 'the coexistence of social and biological rhythms, with the body as the point of contact' and how 'our biological rhythms of sleep, hunger and thirst, excretion and so on are more and more conditioned by the social environment and our working lives' (Elden, 2004: xii). Highmore (2005) points out how Lefebvre and Régulier (1985) suggest that 'only an understanding that can recognize how biological rhythms are orchestrated by linear rhythms and how biology also exceeds this orchestration, is going to be up to the job of comprehending everyday life' (322), particularly in terms of how 'modern routines (public and private) with their linear time patterns are grafted into the rhythms of the body' (323). It is these 'linear time patterns' and 'the rhythms of the body' which are present in Dave's account. Yet there is more at stake in Dave's account.

Blum (2003) writes about the exclusionary and inclusionary nature of the 'hospitality of the city nights' and how there are 'those who can be inspired by the anonymity of nighttime vulnerability' (157). Dave's account is an example of such inspiration. Yet it is also an account that is produced from a position of privilege (see Chapter 3 on socially differentiated walking experiences). The rhythms that resource Dave's account of nighttime walking reflect his experiences as a white, non-disabled, middle-aged man. His enjoyment of walking through the city at night is not a universal one due to the vulnerabilities experienced by different groups of urban pedestrians, which are frequently gendered and racialised (see Kern, 2020 on how race and sexuality influence encounters in public space). Paying closer attention to the rhythms of urban walking makes visible the vulnerabilities and associated inequalities of different walking bodies. Such a focus extends concerns with the vulnerabilities of pedestrians beyond simply considering traffic and road safety but to the ways in which the urban walker experiences vulnerabilities in a social sense. As such, the final section of this chapter considers a 'politics of rhythm' (Mels, 2004: 36) in relation to the socially differentiated nature of the rhythms of walking bodies and the conflicts that can emerge from such differentiated rhythms. For while attention has been paid to walking as a form of everyday political resistance (de Certeau, 1984), far less attention has been paid to the politics associated with the spatio-temporal rhythms of everyday walking practices, particularly in the context of broader concerns with low-carbon mobility.

Conflict, vulnerabilities, and the politics of rhythm

This De Beauvoir Town resident reflects upon her walking patterns in the local area:

> ...once you start pushing a buggy you notice every crack and every raised manhole cover and every really badly done kerb and you tend to sort of start working out the fastest routes to places based on where you've got you know drop downs off the kerb because even quite low kerbs are difficult with a lot of shopping, and you tend to have this sort of mental map.
>
> (interview – Ros, De Beauvoir Town resident, Hackney)

Ros's account is constructed in relation to what it is to walk accompanying a small child ('once you start pushing a buggy') and the spatial constraints imposed by pushing a child's buggy. For example, how 'every crack and every raised manhole cover and every really badly done kerb' are problematic, particularly 'with a lot of shopping'. Ros refers to her 'mental map' that incorporates these restrictions as she negotiates the built environment. Mental mapping has a long history in geography and urban studies (for example see Gould and White, 1974; Lynch, 1960; Tuan, 1975). Many of these writings on mental mapping are based on an understanding that conceives of maps as a cognitive model that guides action. However, subsequent writings problematise this concept by drawing attention to the complexity of wayfinding practices (see Chapter 2 for a detailed discussion of pedestrian wayfinding). For example, Tonkiss (2005) highlights how an individual's 'mental map' emerges from their spatial practices and 'compose a mishmash of landmarks, personal haunts, good guesses and routine paths' which are intimately related to 'the textures of everyday movement in the city, the chance encounters and crosscutting paths of the urban crowd, the tricky and momentary ways in which people make space' (128). It is these very points to which Ros's account is orientated.

However, as Ros refers to the things she notices, how she works out 'the fastest routes', and her knowledge of the built environment, this not only begins to highlight the complexity of how paths are constructed, imagined, and lived but also how spatial practices, such as walking, are also temporal. For as Massey (2001) points out, 'all our lives – are lived in spaces (time-spaces) that are far more complex than you would ever divine from maps of physical mobility' (459). Tonkiss (2005) writes how 'different social actors' in fact 'have quite different "spatial stories" to tell about their routes through the city' (113). Yet the significance of Ros's account also relates to Mels' (2004) contention that any engagement with a 'politics of rhythm' must take account of 'a multiplicity of objective and subjective rhythms of time-space and the part of human practices, hopes and fears in their creation' (36). In other words, differentiated lived experiences are central to understanding the everyday rhythms of contemporary urban life, particularly in relation to inequalities, difference, and everyday power relations (see for example Jirón, 2010; Lager et al., 2016; Schwanen et al., 2012; Spinney, 2010). Ros makes available pedestrian rhythms that highlight her experiences as a mother encumbered not only by journeying with young children but also by the 'equipmentality' of parenting in the form of a buggy laden with shopping. This makes further visible how everyday pedestrian experiences emerge from the multiple ways in which walking is socially and materially co-produced.

This London Fields resident describes her walking patterns in the context of her daily journey to work:

> A lot of people think you can park anywhere with a blue badge, it's not true, it's not true, it's quite restrictive. So that puts me through a lot of anxiety then having to get to somewhere and seeing if I can park there and so on. And I don't want to have to park so far away that it becomes difficult for me to walk. After about a hundred yards or so even with a

stick I start to suffer joint pains and stiffness and things. I find that a bit
nerve wracking particularly if I have to go somewhere I don't know. I
tend to leave very early for things and end up arriving far too early but
that's because I want to have time to circuit, circumnavigate the area.

(interview – Sandra, London Fields resident, Hackney)

Sandra constructs her account in relation to the restrictions she encounters due to being disabled. Sandra's account of these restrictions reflects
the hostile built environments that disabled people confront in their
everyday lives and the challenges they face as they negotiate the city (see
Imrie, 1996, 2012; Middleton and Byles, 2019; Pyer and Tucker, 2017). Yet,
as Sandra describes how her mobility has decreased and she is forced to
increasingly rely on the car, her account also draws attention to the ways
walking can be understood in relation to spatial practices in the city. For
instance, 'wayfinding', or what Ingold (2007) refers to as 'wayfaring', is
a salient aspect of Sandra's account as she re-negotiates her journey to
work on a daily basis so as to avoid parking 'so far away that it becomes
difficult for me [Sandra] to walk'. In explaining how she often leaves 'very
early for things' to increase her chances of finding a convenient parking
space and decreasing the distance she has to walk, Sandra reflects upon
the pain and anxiety she experiences ('puts me through a lot of anxiety';
'suffer joint pains and stiffness and things'; 'a bit nerve wracking'). This
demonstrates how space is both sensually and emotionally apprehended
and what Ingold and Kurttila (2000) refer to as a 'multisensory awareness of the environment' that is key to 'spatial orientation and the coordination of activity' (189). This feature of Sandra's account also draws
attention to the inter-related temporal and spatial dimensions of how she
coordinates her journey and emergent rhythms. However, these inter-
related rhythms are far from smooth due to a series of spatio-temporal
tensions around the timings of Sandra's journeys, the availability of
parking, and how far she has to walk to her destination.

Cresswell (2014) uses the notion of 'friction' as a means of thinking about
how movements are hindered and enabled. He highlights how friction slows
things down but also enables movement and is a 'social phenomenon with
its own politics' (114). It is possible to consider the rhythms and associated
tensions emerging from people's everyday walking practices in similar
terms. In particular, considering the rhythms and 'frictions' of pedestrian
movement not only brings into focus the vulnerabilities of different walking bodies but also how conflict with others emerges as a central concern
for urban walkers. For example, the interview extract below, with another
London Fields resident, is taken from a discussion concerning the potential
fear experienced as a pedestrian:

And I've never been assaulted, I've never actually had a run in. I mean
I've been yelled at, I get yelled out by cyclists or by motorists when they

think I'm in their way. Actually I've been harassed in my life for being with my partner whenever in various cities, in safe places and that's always a threat for us. We got accosted in London Fields by a teenager who just wanted to know why we were laughing and having so much fun together, I think we might have been holding hands or linking arms or something like that. So that for is an underlying threat that we always will have and so I don't distinguish between day and night as my threat so much as that there are just people out there who hate me for who I am and so I carry that around all the time. I don't necessarily look like a lesbian, I don't necessarily look like the object of anybody's hate but I know that I am.

(interview – Lindsey, London Fields resident, Hackney)

In her account, Lindsey makes it explicit that she is a lesbian and how her personal safety has little to do with distinguishing between day and night but 'that there are just people out there who hate me for who I am and so I carry that around all the time'. Lindsey makes reference to the places where herself and her partner have been harassed and threatened, high-lighting the relationship between place and identity (Carter et al., 1993). The incident in London Fields of her and her partner being accosted by a teenager illustrates how they experience their sexuality as 'out-of-place' in this area (see also Browne et al., 2009; Gorman-Murray and Waitt, 2009; Nash and Gorman-Murray, 2014; Valentine, 1993). This is further reflected in how Lindsey explains how she does not think she 'necessarily look[s] like a lesbian' or is the 'object of anybody's hate' but at the same time she knows that she is. Yet, the conflict described in Lindsey's account can also be understood in relation to the rhythms of her and her partner's walking bodies as they hold hands and link arms. The increase in violence and harassment reported by LGBTQ people in both the UK and many other countries has risen sharply in recent years particularly on the streets and in the public spaces of cities. For example, in the UK in 2017, one in five LGBTQ people experienced a hate crime or incident due to sexual orientation and/or gender identity, with four in five people not reporting it to the police (Stonewall, 2017). These figures are likely to be much higher, especially in countries where such data does not exist. For example, South Africa has a lack of official statistics as the law does not classify hate crimes differently from other crimes, while in Botswana, the criminal justice system does not recognise sexual orientation and gender motivated hate crimes, so lacks a reporting mechanism for documenting these forms of violence against LGBTQ people (Iranti, 2019). However, it is not hard to imagine that many of these incidents occur when the different rhythms of walking bodies collide with those that do not fit some misconceived 'norm', leading to conflict of some kind.[1] Such rhythmic conflict not only further highlights the socially differentiated nature of urban walking but also that experiencing walking in the city as an emancipatory and democratic everyday practice is a romanticised vision of walking certainly not experienced by everyone.

Conclusion

Time is a relatively neglected concern across transport research, policy, and practice or an issue that has historically been informed by linear under-standings of time bound up with concerns with speed and efficiency that hinges on universal clock time. In contrast, this chapter shows how time is as an issue of great significance for walking. By engaging with the notion of 'time in events' (Adam, 1990), I have argued that the relationship between walking and time is not only one of clock time passing, but is made up of multiple temporalities which emerge out of and shape people's experiences on foot. These include temporalities such as 'collective times' and 'experien-tial times'. I have drawn upon the 'event' of waiting as particularly signifi-cant for understanding how people experience time and become aware of its duration. The work of Bergson (1911/1998) is productive here to extend this analysis in relation to how he used the notions of 'duration' and 'hesitation' to conceptualise how time is experienced. By engaging with these 'experi-ential times' I have demonstrated the ways in which people's 'sense' of time expands and contracts as they move on foot. This opens up the possibility to do other things in a way which other forms of urban transport are unable to do. These parallel activities include things such as talking on the phone, spending time with family members or friends, or planning the working day ahead. While these may appear seemingly mundane activities, they are significant features of people's everyday lives that are illustrative of how walking allows much more than mere transportation. For, as Solnit (2014) points out, 'it [walking] is an observer's state, cool, withdrawn, with senses sharpened, a good state for anybody who needs to reflect or create' (186). Other emergent temporalities, such as 'collective times' and how decisions to walk are intimately bound up with people's day-to-day routines, are also of relevance here with the diary and interview data illustrating how walking resources individual's and families' everyday life coordination.

The data I have discussed in this chapter reveals the multiple spatiali-ties that are constituted by and practised through walking. Issues emerging from differentiated mobilities illustrate how walking is an inherently spatial practice, connected to a sense of identity. Through this empirical engage-ment, the complexity of the spatio-temporal practices of urban walking, the significance of local knowledge with respect to how pedestrians negotiate the city, and the constantly shifting and unfolding nature of wayfinding on foot are made relevant. Both temporal and spatial concerns are significant resources in people's accounts of their walking practices as they construct their identities and who they are in relation to others. Engaging with people's spatio-temporal experiences on foot matters to understandings of walking, walkability, and the walkable city as it provides a further illustration of the ways in which walking is socially differentiated and even varies with the same person over, with, and through time and space. As such, I have not only considered the notion of rhythm as a productive means for engaging with how time, space, and identity inter-relate as people walk but also as a lens for examining the vulnerabilities and associated inequalities of different walking bodies.

Lefebvre's rhythmanalysis provides a theoretical departure point for such a consideration of rhythm. For, as Buttimer (1976) argues, 'neither geodesic space nor clock/calendar time is appropriate for the measurement of experience', yet 'the notion of rhythm may offer a beginning step toward such a measure' (289). In this chapter, I have examined the rhythms of walking bodies and the conflicts and vulnerabilities relating to differentiated pedestrian rhythms. In particular, I have demonstrated their significance to everyday walking practices and how they relate to both the practice and performance of urban social identities. These bodily rhythms and associated frictions (see Cresswell, 2014) are resources in people's accounts of their walking practices as they construct their identities of who they are in relation to others. Yet, how do these identity politics emerge from the social encounters people experience, or not, on foot? Furthermore, how should we understand and conceptualise the significance of pedestrian encounters to broader concerns with the socially differentiated nature of everyday urban mobilities? It is to these questions I turn in the next chapter.

Note

1. For example, a recent article in *GQ*, the British men's lifestyle magazine, asks 'why do gay men walk so fast?' The author reflects upon his own experiences as a gay man and how 'growing up, people would often tell me that I "walked gay"' (see https://www.gq.com/story/move-im-gay).

References

Adam, B. 1990. *Time and social theory*, Cambridge, Polity.
_____. 1995. *Timewatch: The social analysis of time*, Cambridge, Polity.
Amin, A. & Thrift, N. J. 2002. *Cities: Reimagining the urban*, Cambridge, Polity.
Banister, D. 2011. The trilogy of distance, speed and time. *Journal of Transport Geography*, *19*, 950–959.
Bastian, M. 2017. Liberating clocks: Developing a critical horology to rethink the potential of clock time. *New Formations*, *92*, 41.
Beaumont, M. 2015. *Nightwalking: A nocturnal history of London, Chaucer to Dickens*, London, Verso.
Bell, L. 1998. Public and private meanings in diaries: Researching family and childcare. In: Ribbens, J. & Edwards, R. (eds.) *Feminist dilemmas in qualitative research: Public knowledge and private lives*, London, Sage, 72–86.
Bergson, H. 1991 [1908]. (translated by N. M. Paul and W. S. Palmer). *Matter and memory*, New York, Zone Books.

_____. 1992 [1946]. (translated by M. L. Cunningham Andison). *The creative mind: An introduction to metaphysics*, New York, Citadel.

_____. 1998 [1911]. (translated by N. M. Paul and W. S. Palmer). *Creative evolution*, New York, Dover.

_____. 2001 [1913]. (translated by F. L. Pogson) *Time and free will: An essay on the immediate data of consciousness*, New York, Dover.

Bissell, D. 2007. Animating suspension: Waiting for mobilities. *Mobilities: Translocal Subjectivities: Mobility, Connection, Emotion, 2*, 277–298.

Bissell, D. 2009. Conceptualising differently-mobile passengers: Geographies of everyday encumbrance in the railway station. *Social & Cultural Geography, 10*, 173–19.

Bissell, D. 2018. *Transit Life: How commuting is transforming our cities*, Cambridge, MA, MIT Press.

Blum, A. 2003. *The imaginative structure of the city*, Montreal, Ithaca, McGill-Queen's University Press.

Browne, K., Lim, J. & Brown, G. (eds.). 2009. *Geographies of sexualities: Theory, practices and politics*, Aldershot, Ashgate Publishing.

Buttimer, A. 1976. Grasping the dynamism of lifeworld. *Annals of the Association of American Geographers, 66*, 277–292.

Carlson, J. A., Saelens, B. E., Kerr, J., Schipperijn, J., Conway, T. L., Frank, L. D., Chapman, J. E., Glanz, K., Cain, K. L. & Sallis, J. F. 2015. Association between neighborhood walkability and GPS-measured walking, bicycling and vehicle time in adolescents. *Health & Place, 32*, 1–7.

Carter, E., Donald, J. & Squires, J. 1993. *Space and place: Theories of identity and location*, London, Lawrence, Wishart.

City of Melbourne. 2014. *Walking Plan* [available at: https://s3.ap-southeast-2.amazonaws.com/hdp.au.prod.app.com-participate.files/4114/3890/9931/Walking_Plan_full_version.pdf].

Clayton, W., Jain, J. & Parkhurst, G. 2017. An ideal journey: Making bus travel desirable. *Mobilities: Curated Issue: Transport Mobilities, 12*, 706–725.

Crang, M. 2001. Rhythms of the city: Temporalised space and motion. In: May, J. & Thrift, N. J. (eds.) *Timespace: Geographies of temporality*, London, Routledge, 187–207.

Cresswell, T. 2014. Mobilities III: Moving on. *Progress in Human Geography, 38*, 712–721.

Crow, G. & Heath, S. 2002. *Social conceptions of time: Structure and process in work and everyday life*, Basingstoke, Palgrave Macmillan.

Davies, K. 2001. Reflections over timespace. In: May, J. & Thrift, N. J. (eds.) *Timespace: Geographies of temporality*, London, Routledge, 133–148.

De Certeau, M. 1984. *The practice of everyday life*, Berkeley, London, University of California Press.

Department for Transport. 2018. Transport statistics Great Britain 2018 [available at: https://assets.publishing.service.gov.uk/ government/uploads/system/uploads/attachment_data/file/787488/tsgb-2018-report-summaries.pdf].

Durkheim, É. 1976 [1915]. (translated by J. W. Swain). *The elementary forms of the religious life*, London, Allen and Unwin.

Elden, S. 2004. (translated by Rhythmanalysis: An introduction. In: Elden, S., Moore, G. & Lefebvre, H. (eds.) *Rhythmanalysis: Space, time, and everyday life*, London, Continuum, VII–XV.

Gell, A. 1992. *The anthropology of time: Cultural constructions of temporal maps and images*, Oxford, Berg.

Glennie, P. & Thrift, N. J. 2009. *Shaping the day: A history of timekeeping in England and Wales, 1300–1800*, Oxford, Oxford University Press.

Goodman, R. 2001. A traveler in time: Understanding deterrents to walking to work. *World Transport Policy and Practice*, 7, 50–54.

Gorman-Murray, A. & Waitt, G. 2009. Queer-friendly neighbourhoods: Interrogating social cohesion across sexual difference in two Australian neighbourhoods. *Environment and Planning A*, 41, 2855–2873.

Gould, P. & White, W. R. 1974. *Mental maps*, Harmondsworth, Penguin.

Grönland, B. 1998. *Lefebvre's Rhythmanalysis*, Copenhagen, Seminar with the Center for Cross-disciplinary Urban Studies at the School of Architecture.

Hägerstrand, T. 1967. *Innovation diffusion as a spatial process*, Chicago, London, University of Chicago Press.

_____. 1975. Space, time and human conditions. *Dynamic Allocation of Urban Space*, 3.

_____.1977. The geographers' contribution to regional policy: The case of Sweden. In: *Geographic humanism, analysis and social action: A half century of geography at Michigan. Michigan geographical publications*. No. 17. Ann Arbor, MI, University of Michigan Press, 329–346.

Hagman, O. 2006. Morning queues and parking problems. On the broken promises of the automobile. *Mobilities*, 1, 63–74.

Harris, P., Lewis, J. & Adam, B. 2004. Time, sustainable transport and the politics of speed. *World Transport Policy and Practice*, 10, 5–11.

Highmore, B. 2005. *Cityscapes: Cultural readings in the material and symbolic city*, Basingstoke, Palgrave Macmillan.

Imrie, R. 1996. *Disability and the city: International perspectives*, London, Paul Chapman.

_____. 2012. Auto-disabilities: The case of shared space environments. *Environment and Planning A: Economy and Space*, 44, 2260–2277.

Ingold, T. 2007. *Lines: a brief history*, London, Routledge.

Ingold, T. & Kurttila, T. 2000. Perceiving the environment in Finnish Lapland. *Body & Society*, 6, 183–196.

Iranti. 2019. Data collection and reporting on violence perpetrated against LGBTQI persons in Botswana, Kenya, Malawi, South Africa and Uganda. Arcus Foundation [available at: https://www.iranti.org.za/wp-content/uploads/2020/03/Violence-Report_36482.pdf].

Jain, J. 2011. The classy coach commute. *Journal of Transport Geography*, 19, 1017–1022.

Jeffrey, C. 2010. *Timepass: Youth, class, and the politics of waiting in India*, Stanford, Stanford University Press.

Jirón, P. 2010. Mobile borders in urban daily mobility practices in Santiago de Chile. *International Political Sociology*, 4, 66–79.

Joshi, M. S. & Senior, V. 1998. Journey to work: The potential for modal shift? *Health Education Journal*, 57, 212–223.

Kern, L. 2020. *Feminist City: Claiming space in a man-made world*, London, Verso.

Lager, D., Van Hoven, B. & Huigen, P. P. P. 2016. Rhythms, ageing and neighbourhoods. *Environment and Planning A*, 48, 1565–1580.

Latham, A. 2003. Research, performance, and doing human geography: Some reflections on the diary-photograph, diary-interview method. *Environment and Planning A*, 35, 1993–2017.

Latham, A. & Layton, J. 2019. Social infrastructure and the public life of cities: Studying urban sociality and public spaces. *Geography Compass*, 13.

Latour, B. 2005. Trains of thought: The fifth dimension of time and its fabrication. In: Perret-Clermont, A. N. (ed.) *Thinking time: A multidisciplinary perspective on time*, Cambridge, MA, Hogrefe and Huber, 173–187.

Laurier, E., Lorimer, H., Brown, B., Jones, O., Juhlin, O., Noble, A., Perry, M., Pica, D., Sormani, P., Strebel, I., Swan, L., Taylor, A. S., Watts, L. & Weilenmann, A. 2008. Driving and 'passengering': Notes on the ordinary organization of car travel. *Mobilities*, *3*, 1–23.

Lefebvre, H. 2004. *Rhythmanalysis: Space, time, and everyday life*, London, Continuum.

Lefebvre, H. & Régulier, C. 1985. Le projet rythmanalytique. *Communications*, *41*, 191–199.

Levine, M. 2003. Times, theories and practices in social psychology. *Theory & Psychology*, *13*, 53–72.

Lynch, K. 1960. *The image of the city*, Cambridge, MA, MIT.

Lyon, D. 2016. Doing audio-visual montage to explore time and space: The everyday rhythms of Billingsgate Fish Market. *Sociological Research Online*, *21*, 57–68.

_____. 2018. *What is rhythmanalysis?* London, Bloomsbury Publishing.

Lyons, G. & Urry, J. 2005. Travel time use in the information age. *Transportation Research. Part A, Policy and Practice*, *39*, 257–276.

Massey, D. 2005. *For space*, London, Sage.

Massey, D. B. 2001. Living in Wythenshawe. In: Borden, I., Kerr, J., Rendell, J. & Pivaro, A. (eds.) *The unknown city: Contesting architecture and social space*, Cambridge, MA, London, MIT, 459–475.

Matos Wunderlich, F. 2008. Walking and rhythmicity: Sensing urban space. *Journal of Urban Design*, *13*, 125–139.

Mels, T. 2004. Lineages of a geography of rhythms. In: Mels, T. (ed.) *Reanimating places: A geography of rhythms*, Aldershot, Ashgate, 3–42.

Middleton, D. & Brown, S. D. 2005. *The social psychology of experience: Studies in remembering and forgetting*, London, Sage.

Middleton, J. & Byles, H. 2019. Interdependent temporalities and the everyday mobilities of visually impaired young people. *Geoforum*, *102*, 76–85.

Nash, C. J. & Gorman-Murray, A. 2014. LGBT neighbourhoods and 'new mobilities': Towards understanding transformations in sexual and gendered urban landscapes. *International Journal of Urban and Regional Research*, *38*, 756.

Pae, G. & Akar, G. 2020. Effects of walking on self-assessed health status: Links between walking, trip purposes and health. *Journal of Transport & Health*, 18.

Pyer, M. & Tucker, F. 2017. 'With us, we, like, physically can't': Transport, mobility and the leisure experiences of teenage wheelchair users. *Mobilities: Curated Issue: Slow Mobilities*, *12*, 36–52.

Saevereid, H. A., Schnohr, P. & Prescott, E. 2014. Speed and duration of walking and other leisure time physical activity and the risk of heart failure: A prospective cohort study from the Copenhagen City eHeart Study. *PLoS ONE*, *9*, e89909.

Sahlqvist, S., Song, Y. & Ogilvie, D. 2012. Is active travel associated with greater physical activity? The contribution of commuting and non-commuting active travel to total physical activity in adults. *Preventive Medicine*, *55*, 206–211.

Schwanen, T., Hardill, I. & Lucas, S. 2012. Spatialities of ageing: The co-construction and co-evolution of old age and space. *Geoforum*, *43*, 1291–1295.

Sheller, M. & Urry, J. 2006. The new mobilities paradigm. *Environment and Planning A*, *38*, 207–226.

Simonsen, K. 2004. Spatiality, temporality and the construction of the city. In: Bærenholdt, J. O. & Simonsen, K. (eds.) *Space odysseys: Spatiality and social relations in the 21st century*, Aldershot, Ashgate, 43–62.

Singleton, P. A. 2019. Discussing the "positive utilities" of autonomous vehicles: Will travellers really use their time productively? *Transport Reviews*, *39*, 50–65.

Smith, R. & Hall, T. 2012. No time out: Mobility, rhythmicity and urban patrol in the twenty-four hour city. *The Sociological Review*, *60*, 89–108.

Solnit, R. 2014. *Wanderlust: A history of walking*, London, Verso.

Spinney, J. 2010. Performing resistance? Re-reading practices of urban cycling on London's South Bank. *Environment and Planning A*, *42*, 2914–2937.

Stonewall. 2017. LGBT in Britain: Hate crime and discrimination, YouGov [available at: https://www.stonewall.org.uk/lgbt-britain-hate-crime-and-discrimination].

Tang, J., Zhen, F., Cao, J. & Mokhtarian, P. 2018. How do passengers use travel time? A case study of Shanghai–Nanjing high speed rail. *Planning – Policy – Research – Practice*, *45*, 451–477.

Thrift, N. 1977. *An introduction to time geography*, Norwich, Geo Abstracts.

Tonkiss, F. 2005. *Space, the city and social theory: Social relations and urban forms*, Cambridge, Polity.

Transport for London. 2020. *Boost for walking as TfL launches Central London Footways created by London Living Streets and Urban Good*. 17th September [available at: https://tfl.gov.uk/info-for/media/press-releases/2020/september/boost-for-walking-as-transport-for-london-launches-central-london-footways-created-by-london-living-streets-and-urban-good].

Tranter, P. & Tolley, R. 2020. *Slow cities: Conquering our speed addiction for health and sustainability*, Oxford, Elsevier.

Tuan, Y.-F. 1975. Images and mental maps. *Annals of the Association of American Geographers*, *65*, 205–212.

Valentine, G. 1993. (Hetero) sexing space: Lesbian perceptions and experiences of everyday spaces. *Environment and Planning D: Society and Space*, *11*, 395–413.

Wardman, M. & Lyons, G. 2016. The digital revolution and worthwhile use of travel time: Implications for appraisal and forecasting. *Planning – Policy – Research – Practice*, *43*, 507–530.

Watts, L. & Urry, J. 2008. Moving methods, travelling times. *Environment and Planning D: Society and Space*, *26*, 860–874.

5   Pedestrian politics of (non)encounters

In 2008, Ingold and Vergunst observed, 'Not only, then, do we walk because we are social beings, we are also social beings because we walk. That walking is social may seem obvious, although it is all the more remarkable, in this light, that social scientists have devoted so little attention to it' (2). Yet, despite the passing of nearly 15 years, there still remains very little explicit attention to the socialites of urban walking. In attempting to address this impasse and engage with the complexities of the social dimensions of walking I have previously argued that considering the nature of everyday pedestrian encounters in the broader context of the frequently contested use of urban space on foot opens up new ways of thinking about everyday urban politics (Middleton, 2018). In particular, it enables the concept of the 'right to the city' to extend beyond political slogans and an over-emphasis on social movements towards a focus on everyday micropolitics and practical policy concerns. This chapter seeks to develop this work through further examining notions of the right to the city in the context of how it is claimed and embodied (Duff, 2017), and more specifically how the 'right to mobility' relates to the 'right to encounters', in the everyday experiences of urban walking. Who has the right to initiate, perform, and resist pedestrian encounters?

In particular I will draw upon Kärrholm et al.'s (2017) invitation to reflect on how different sorts of walking practices are assembled to co-exist (or not) in order to gain new insights into the subtle power relations of urban mobility in public space. In doing so, I consider the significance of both encounter and non-encounter to everyday walking practices. For while walking through the city involves the unfolding of multiple encounters, it is also comprised of the encounters which do not emerge: avoiding eye contact with street vendors or zigzagging to dodge the forceful, yet always cheery, approach of charity 'chuggers'.[1] Simmel (1971) writes of the blasé indifference of urban dwellers as they negotiate the city on a daily basis, while Goffman (1963) refers to 'civil inattention' where strangers are registered but not acknowledged. For Kern (2020) being able to be alone is what constitutes a 'successful city' (89). She goes on to explain that: 'The extent to which violations of women's personal space via touch, words, or other infringements are tolerated and even encouraged in the city is a good as measure as any for me of how far away we actually are from the sociable – and feminist – city of spontaneous encounters' (ibid). Yet, what I want to argue here is that pedestrian non-encounters warrant further attention. In particular, I will challenge the ways in which pedestrian encounters are frequently framed as positive or, at the very least, benign. Such assumptions cut across both academic and policy spheres, yet the reality is a complexity which is rarely engaged with. As such, I explore this complexity through focusing on two different forms of urban pedestrian patrolling. I demonstrate how pedestrian encounters and non-encounters are of particular significance for understanding the negotiated politics of everyday life in the city. In this context, urban patrols are understood as repetitive pedestrian movements through the city in response to the needs of certain groups (see Hall and Smith, 2013). Street Pastors[2] are an inter-denominational church initiative aimed at providing care in city centres, particularly in close proximity to spaces of the night-time economy. The SockMob[3] are a volunteer network who participate in regular weekly walks in central London using socks as an ice-breaker to initiate conversations with people living on the streets.

I argue that both Street Pastor and Sock Mob pedestrian urban patrols are part and parcel of a series of unequal outcomes and experiences that relate more broadly to the politics of pedestrian encounters. Through an examination of the night-time patrolling activities of each group I examine the moralities associated with their pedestrian encounters. A series of moral codes emerge from the analysis of in-depth interview data and participant observation in relation to notions of appropriate conduct and ideas of who or what belongs in the spaces they patrol. These emergent moralities also inform judgments as to who or what is deemed 'vulnerable' and an appropriate recipient of the forms of care being offered (see also Spinney, 2015, on citizenship as a moral accomplishment in the context of HGV drivers). These judgments form part of the unfolding of both pedestrian encounters and non-encounters that have wider resonance to how we understand the subtle power relations of urban mobility in public space in terms of how people appropriate space in certain ways and the moralities that frequently inform these everyday practices. I examine how these distinct forms of pedestrian patrolling contribute, in different ways, to established orders of governance. In what follows, I argue that walking is not an innocent urban practice and the power relations associated with the everyday encounters that do, and do not, unfold on foot matter to how walking in the city is understood. Yet where should we start in beginning to understand and conceptualise the 'in-between' spaces of pedestrian encounters and their potential to produce particular affective atmospheres?

Encounters, the 'good city', and the 'right to mobility'

Recent years have seen a growing interest in urban encounters. In contrast to the dystopic view of urban life, characterised by a focus on inequality and exclusion, that dominated urban studies through much of the 1990s and early 2000s (Baiocchi, 2001; Mohan, 2000; Wacquant, 1999; Wilson, 1997), a number of scholars have explored the potential of what Amin (2006) has referred to as the 'good city'. Underpinning these arguments is recognition that, despite growing multiplicity and inequality, a certain 'low level sociability' is still apparent in our cities through which people learn to live with difference. Though often overlooked, for Thrift (2005), the mundane friendliness that characterises many of the everyday interactions between strangers represents a kind of 'base-line democracy' from which a more progressive urban politics might be built. Situated within these broad concerns, from a range of multiple perspectives, have been calls to pay more attention to the micro-scale complexity of urban encounters.

Hall and Smith (2014) build upon and extend writings on urban repair in relation to the 'good city' through a focus on street cleaning and urban outreach work with street homelessness. They celebrate the notion of envisioning 'hope' in cities by highlighting the significance of both physical and social repair. However, they urge for a greater acknowledgement of the politics of repair and 'enquiry into the accomplishment and politics of street repair' (13). Valentine (2008) also welcomes the more positive focus of work on notions of the 'good city', yet raises an important note of caution in suggesting that:

> Much of the writing that is associated with what might be regarded as a 'cosmopolitan turn' … celebrates the potential for the forging of new … ways of living together with difference *but* without actually spelling out how this is being, or might be, achieved in practice. (324)

While Hall and Smith (2014) note the abstract nature of the arguments surrounding Thrift's (2005) notions of urban care and repair, Valentine (2008) has concerns beyond a lack of empirical work examining the practical dynamics of urban encounters in the 'good city'. First, she suggests we need to differentiate more carefully between different forms of encounter. As she notes, contact with difference is as likely to foster intolerance and resentment as any newfound respect for difference. Second, Valentine suggests we need to distinguish between basic civility and tolerance. She argues this actually signals an unequal relationship of power between those doing the tolerating and those being tolerated and what she calls more 'meaningful moments of contact' that might engender a genuine respect for difference on its own terms. And third, she urges us to consider how such moments might subsequently be 'scaled up' beyond the moment of encounter itself so as to foster a wider sense of the good city.

Wilson (2017) warns that while the notion of encounter has been widely used in research on urban diversity and socio-cultural difference, it is at risk of being under-theorised. In pulling together a diverse range of engagements with notions of encounter, she highlights the dangers of it becoming an 'empty' concept 'which undermines the critical and analytical force of work that engages it as a key site of scholarly interest' (2). She also raises two points of caution in relation to Valentine's account of meaningful encounters. The first relates to notions of 'actual values and belief' not being situated as 'separate and formed in isolation from encounters' (460). Wilson argues that this conceptual move avoids values and beliefs being rendered 'fixed, stable and clearly defined' (460). Second, she highlights the importance of questioning who an encounter is meaningful for and the politics associated with who exactly is making such a value judgment. Some of the most notable work in urban studies on forms of non-encounter is research concerned with the spatial segregation of residents in middle-class urban enclaves (Bulut, 2018; Low, 2001; Morgan, 2013). For example, Schuermans' (2016) detailed study of enclaves in Cape Town examines the mobilities both within and between enclaves. He argues that much existing work fails to fully understand the effects of enclave urbanism due to a narrow and overly sedentarist consideration of the nature of urban encounters. He suggests the need for attention being given to 'both sedentary and mobile encounters inside and in-between various kinds of enclaves' (184).

Concerns with people's rights to move in, through, across, and between different places are fundamental to understandings of everyday urban mobility, because 'the capacity to move is central to what it is to be a citizen' (Cresswell, 2009: 110). The links between walking encounters and urban sociability can be thought about more closely in relation to the right to the city and the everyday tactics of urban pedestrians. Much has been written concerning the right to the city, with some of the most influential work emerging from the work of Lefebvre (1970, 1991). For Lefebvre, citizenship and political belonging emerge from the notion of the right to the city in the form of inhabitance as opposed to formal citizenship status. Purcell (2003) argues that these 'rights' can be considered in two inter-related categories. The first is the right to appropriate space in terms of inhabitants having the right to occupy, use, work, live, etc., in specific city spaces. The second is the right to participate in decision-making at various political scales in the production of urban space. It is the former category that has most relevance for the context of this chapter. As I discussed in Chapter 3, the politics of urban walking have been more directly engaged with in the seminal work of de Certeau (1984) in relation to pedestrian 'tactics' as a form of resistance to the 'strategies' of urban planners.

Engagements with, and critiques of, both Lefebvre and de Certeau are multiple and extensive with the notion of a right to the city having been mobilised in different ways. Within academic literature the concept has been drawn upon in contexts ranging from social movements and protest (De Souza, 2010), segregation and gentrification (Smith, 1996), race and racial identities (McCann, 1999), gendered and feminist perspectives on new forms of citizenship (Fenster, 2005), mobility (Attoh, 2012; Verlinghieri and Venturini, 2018), and surveillance and social control (Hubbard, 2004). Harvey (2008) proposes that the struggles of the urban crisis should be addressed by adopting right to the city as a 'working slogan and political ideal' (40) as a means of contributing to a broader social movement necessary for the urban dispossessed to take back control. For Mitchell (2003), the key question is who has the right to the city and its public spaces such as parks and squares. He questions the 'tenuous nature of Lefebvre's "right to the city"' (5) in arguing that any rights are dependent on concerns with public space.

More recently, Purcell (2013) highlights how the meanings of the right to the city are 'increasingly indistinct' (141). He moves on to suggest that through a close reading of Lefebvre's original text is a radical understanding of the right to the city that is an 'essential element of a wider political struggle for revolution' (142). In doing so he identifies a gap between Lefebvre's radical conception of the right to the city and contemporary urban initiatives concerned with the 'struggle to augment the rights of urban inhabitants against property rights of owners' (142). Merrifield (2011) makes a related argument in pointing to the need to re-coup the radical potential of the right to the city by moving beyond it to a 'politics of encounter' which is not about asking for rights but 'just act[ing" (479). He concludes that this move 'is potentially more empowering because it is politically and geographically more inclusive' (479). Verlinghieri and Venturini (2018) connect concerns with right to the city with the right to mobility. They argue that while the concept of right to the city has received little explicit attention in transport geography, the mobility turn has seen work focusing on the right to mobility understood as 'expressing the right to move in urban space, to access places and opportunities, but also the right to stay still' (127). They suggest that the concept of the right to mobility enriches understandings of the right to the city in relation to the significant role that mobilities have in the 'production of urban processes' (127) such as access to public space and housing.

Concerns with the right to mobility can be situated within a broader set of issues relating to transport and mobility justice. Verlinghieri and Schwanen (2020) point to three broad ways in which transport and mobility justice has been engaged with and theorised. The first is in relation to transport equity in terms of 'who' is disadvantaged in terms of transport access and distribution (see for example Lucas and Jones, 2012, on the sociodemographic distributional effects of transport policy). The second is where the focus has shifted to concerns with transport justice as 'ongoing process, power relations and struggles over praxis, meaning and values that are actively shaped by the places and spatial configurations as part of which they unfold' (Verlinghieri and Schwanen, 2020). Critical race and gender scholarship, and also the writings of Lefebvre and Harvey on right to the city, has informed much of this work that interrogates the oppression and dispossession experienced by different groups as a direct result of transit and transport systems (see for example Castañeda, 2020; Golub and Martens, 2014; Sukaryavichute and Prytherch, 2018). A focus on mobility justice constitutes the third shift they outline. In particular, the work of Sheller (2018) is highlighted and her conceptualisation of mobility justice that centres concerns with how power and inequality shape mobilities in addition to how other actors, such as the state, govern and control mobility. Smeds et al. (2020) contend that such an engagement with mobility justice facilitates thinking beyond distributive justice while providing, in contrast to work on transport equity, a more nuanced understanding of social difference that takes account of the socially constructed nature of social groups/ categories.

However, critiques of such a notion of mobility justice have also emerged. For example, Verlinghieri and Schwanen (2020) raise questions around how academic research and planning might contribute to Sheller's agenda and what are appropriate methodologies and approaches to adopt. While Davidson (2020), proposing a rethinking of 'radicality within geographies of mobility' (2), contends that Sheller's work is still overly concerned with distributive justice. Yet, despite such critiques, concerns with mobility justice allow thinking to move away from primary concerns with differential accessibility and distributive justice to making visible the socially differentiated and embodied nature of mobility. The remainder of this chapter engages with concerns of mobility justice by extending notions of the right to the city and the right to mobility through examining the right to encounters. This focus makes apparent the power relations that unfold from how people appropriate urban space on foot while challenging the ways in which pedestrian social encounters are frequently romanticised and underpinned by a series of positive assumptions. Engaging with this complexity and the frequently contested use of urban space on foot opens up new ways of thinking about inequalities emerging from everyday urban mobilities.

I illustrate these concerns through focusing on the dynamics, and associated subtle power relations, of pedestrian encounters emerging from the urban patrols of Street Pastors and London's Sock Mob. Drawing on ethnographic research, I explore the work of both organisations by paying specific attention to the interactions that unfold with the people they seek out or meet on their walking routes. In particular, I highlight the significance of the mobile nature of their night-time patrolling activities and the specific role of walking. It is through this focus on the micropolitics of pedestrian encounters that I seek to respond to calls, such as those made by Hall and Smith (2014), to pay closer attention to the accomplishment and politics of street repair in the broader context of urban encounters in the contemporary city (Darling and Wilson, 2016; Thrift, 2005; Valentine, 2008). Through the chapter I attend to how these forms of urban pedestrian patrols are accessories to 'unequal outcomes and experiences' (Hall and Smith, 2014: 505). I do so by addressing specific concerns in each group that include notions of what constitutes appropriate conduct; ideas of belonging and who/what are considered 'vulnerable'; and how certain moral codes both affect and emerge out of pedestrian encounters.

I examine the ways in which the walking activities of each group, such as the routines and rhythms of pedestrian patrols, afford the possibilities for encounters to both occur and or/and be resisted. I argue that these pedestrian practices are fundamental to understanding the significance of the right to encounters, and non-encounters, and to the power relations at stake in terms of how people appropriate space as they move through the city on foot. In the next section I provide some background context to each group and some reflections upon the blurred boundaries between my role as a researcher and volunteer.

Street Pastors

I first became aware of Street Pastors following a chance conversation while teaching on a first-year undergraduate geography fieldtrip to the city of Bath in the UK. Late one evening I was returning to our accommodation with several colleagues when we saw in the distance a young woman who was visibly drunk and struggling to walk along the street. As we got closer, two friends appeared and helped her into a taxi before any of us were able to offer assistance. Yet, I could not help begin to imagine the types of encounters that might have unfolded had she been on her own without her friends there to help her. Remembering far too many similar situations as a student myself, I commented to a colleague that it concerned me how common an occurrence this is in city centres up and down the UK and the vulnerable position so many young people find themselves in on a night out. It was at this point that he began describing the work of Street Pastors, an interdenominational Christian organisation of volunteers, who position themselves as filling a void in urban service provision through providing 'on-street care' in the public spaces of the night-time economy.

The first patrols were established in London in 2003 and now operate in over 250 urban locations that range from large cities to market towns in the UK. As the conversation with my colleague, Richard Yarwood, continued it became clear that the activities of Street Pastors intersected with both our individual research interests: myself in relation to everyday urban mobilities and Richard with respect to the geographies of policing and patrolling activities (Paasche et al., 2014; Yarwood, 2007). Upon our return home we developed, in dialogue with the Ascension Trust,[4] a pilot study that focused on Street Pastors as a distinct form of faith-based volunteering, which then developed into a national study of Street Pastors. The research[5] comprised a national survey of groups, interviews with key actors, and a series of mobile participant observations and interviews with Street Pastor groups in a range of urban localities that included London, other large cities, and market towns across the UK. In what follows, I examine the significance of walking to the night-time activities of Street Pastors in terms of how their pedestrian practices are central to the forms of encounters they seek out in their urban patrols.

The Sock Mob

The Sock Mob was started by a group of friends in 2003 who took to walking the streets of central London once or twice a week as a means of engaging with, and offering companionship to, the homeless people they encountered. They became known as the Sock Mob due to the socks they frequently handed out as an icebreaker/initiator of social contact. Since the group's inception they have been clear to position themselves as 'not a charity, or formal group of any kind' but 'a group or similarly-minded people with a desire to put humane principles into everyday, effective action' (Sock Mob, 2018). Though Sock Mobbers will very often pass gifts, such as socks or small items of food or toiletries, to the homeless people they encounter, these gifts are understood as a hook or bridge into a conversation, rather than the main purpose of the encounter itself. In contrast to most Soup Runs, the Sock Mob is also explicit in not having any formal kind of guiding ethos or mission statement and noticeably overtly distances itself from any religious affiliation. They claim that 'this informal structure is central to the ethos of our group, as it keeps us free of any externally legitimising constraints that might otherwise inhibit the intimacy, deep-rooted trust and free-thinking spirit inherent to what we do' (ibid). Mobbers also tend to distance themselves from any self-identification as a volunteer, describing the Sock Mob as a social group that includes homeless people rather than a form of voluntarism or charity. Furthermore, though individual Mobbers may end up sharing information about local services with homeless people if asked, unlike professional outreach workers, the encounters themselves are not primarily centred around a discussion of a person's service needs.

Therefore, in contrast to charitable and voluntary organisations, the Sock Mob describes itself as a 'Meet Up Group'. Meet Up Groups are hosted on the web, and structured around the shared interests of their members that can include anything from local history, to cinema or wine tasting. Members communicate via a message board and arrange their own physical Meet Ups. In Sock Mob's case, Meets Ups take the form of a Mobbing, which involves small groups of anything between three and six people meeting up each week to walk around a different part of central London for a couple of hours, engaging street homeless people in conversation. Since its inception Sock Mob has attracted more than 4000 new members with some taking part only once, while others attend Mobbings on a more regular, weekly, or monthly basis.

My initial involvement with the group followed on from completing the fieldwork with the Street Pastors. During the patrols I had accompanied the Street Pastors on I frequently became frustrated following the safeguarding/risk assessment protocols to remain at a distance and not get involved with any of the interactions they had with people they encountered. These rules were not only difficult to abide to for practical reasons (many people took little notice of my 'observer' bib and engaged me in conversations) but were also frustrating, as many of the things I witnessed compelled me to offer care and/or assistance. My experiences of Street Pastor patrols left me with a sense of wanting to become involved with something that attempted to address some of the situations, and particularly inequalities, I had witnessed on the streets at night. However, being a non-Christian meant I would never be able to become a Street Pastor.[6]

I was aware of the Sock Mob through a friend who had been involved in their initial set up. A move back to London prompted me to begin to participate in Mobbings. A chance conversation with the geographer Jon May also led to his involvement with the group. Jon has longstanding research interests in the geographies of homelessness (see Cloke et al., 2010; May and Cloke, 2014), and together with my research on everyday urban mobilities, we became increasingly interested in the politics of the patrolling activities of the group. As it became clear that our participation had begun to morph into research we, with the consent of other Mobbers, began developing a research project on the weekly mobbings.[7] What follows is based on field-work undertaken between 2011 and 2013 emerging from my initial involvement as a member of the group. At the time of writing, the group is still active with regular patrols organised via the Meetup website. Our research was primarily in the form of participant observation and involved joining up with a number of different Mobbings as they patrolled different parts of central London. Our observations of these encounters with homeless people living on the streets, and our own experiences of them, were recorded in detailed field diaries. This 'go along' method (Kusenbach, 2003) enabled us to observe the experiences of other Sock Mobbers while providing the opportunity for informal discussions emerging from events that unfolded during the evening patrol.

Night walking

Across much academic research, policy, and practice, day-time is the dominant discourse. There are of course notable exceptions where concerns with urban nights have attracted attention, with the nocturnal drawn upon as a lens to further understand urban life (see for example Edensor, 2015; Shaw, 2018; Wilkinson, 2017). Despite what they refer to as the nyctalopia ('nightblindness') of urban studies, Van Liempt et al. (2015) highlight four overlapping foci of research on the urban night: 'the changing meanings and experiences of urban darkness and nights, the evolution of urban night-time economies, the intensification of regulation and the dynamics in practices of going out' (415). Among, and alongside, such work there has also been a specific focus on night-time mobilities. For example, Plyushteva and Boussauw (2020) examine gendered night mobilities in the context of Sofia's night bus network while questioning the broader inclusiveness of night transport services. In doing so, they pose pressing questions around the inclusion and exclusion processes at play in nocturnal cities. While Smeds et al. (2020) focus on the policy discourses around night-time mobilities in London in relation to concerns with mobility justice. They argue for the need for more critical perspectives addressing the politics of whether and how mobility difference is accounted and planned for in transport policy. I will subsequently return to the issue of mobility justice both in this chapter and in Chapter 7.

In the context of urban walking, in much research and policy there is frequently little differentiation between day- and night-time walking beyond discussions of fear of crime and issues such as street lighting. As I discussed in Chapter 4, the rhythms of night walking are distinctive yet frequently overlooked in most considerations of urban walking practices in relation to health, active travel, low-carbon mobilities, and 'liveable' cities. This sits in contrast to more literary and philosophical engagements with urban pedestrian practices and concerns with the 'night walker'. In the book *Dark matters: A manifesto for the nocturnal city*, Dunn (2016) invites the reader to experience the materiality, textures, smells, sights, and sounds of the city differently through walking through the city at night. The analytical and theoretical narrative of the book is interspersed with descriptions of nocturnal walks Dunn has repeatedly undertaken in his home city of Manchester, UK. Beaumont (2015) reminds us that there is nothing new about people writing about their experiences of walking the city streets at night. In his exploration of the concept of nightwalking, he traces the accounts of the night through the writings of British authors such as Shakespeare, Dunton, Wordsworth, Blake, and Dickens. While Beaumont acknowledges the historical absence of the voices of women or those who walked the streets homeless and impoverished, what unites both historical and contemporary accounts of nightwalking is a pre-occupation with the solitary urban walker. To date, there has been little attention on the group dynamics of walking at night. In what follows, I attend to such an omission through focusing on the Street Pastors and SockMob and the group dynamics as they patrol on foot at night. It is through such a focus that the significance of walking to their patrolling activities, and aims of the group more broadly in the seeking out of encounters, become evident.

Seeking encounters: patrolling routes

The purposive seeking out of encounters is central to the activities of each organisation with this being underpinned by a wider discourse positioning social encounters as an inherently positive aspect of contemporary urban life. Not surprisingly patrolling routes are of particular significance for how the encounters they seek out on foot unfold. For the Street Pastors, their patrols occupy a distinct set of time-space routines based around two-hour shifts punctuated by breaks back at a central base established in a church or community centre. The base itself is an important fixed space in the mobility of the patrol, providing a start and end point, and providing a place for refuge and prayer. Each team follows a set route that has usually been planned with the police to ensure the safety and efficiency of the work of Street Pastors. Sometimes only streets covered by closed-circuit television are patrolled. This is partly for reasons of safety but also because CCTV operators will direct Street Pastors to incidents that are suited to their skills. Once on the streets, however, police and Pastors maintain their own patrol patterns and activities. Although they occasionally meet to exchange pleasantries and information, both maintain some distance as a means of demonstrating their organisational independence. Overall, the geographies of these patrols reflect that Street Pastors have been enrolled into wider, secular networks of city management and surveillance. This is reflected in the interview extract below:

> ...the routes were chosen for us but we developed them. The routes mean that roughly every 20 minutes we'll be round again. I learnt that if you do consistent routes, people get to know that you'll be round again, and that's what we operate. We can be found, that's why CCTV can call me in December last year and say that a WPC [Woman Police Constable] needs your help, because they could see me, they knew where to find my teams. And we were available instantly. So it's proven to be a reliable and conclusive method of patrolling. So we're not just willy-nilly where shall we go tonight? It's all set.
>
> (Simon, Street Pastor group leader, large city)

Simon positions his account of patrolling routes in relation to the professional nature in which they develop. For example, he emphasises the training in which he learnt about the importance of the consistency of routes and how they are 'a reliable and conclusive method of patrolling' as opposed to being 'just willy-nilly'. He also stresses the professional conduct of the teams patrolling these routes in relation to the other agencies they interact with, such as CCTV operators and the police, and how they are led by team leaders and can be 'available instantly'. Simon's account articulates how patrols aim to follow a routine that maximises the chances of encountering people on the streets and is informed by training and other professional agencies. In other words, Simon's account draws attention to how Street Pastors use their pedestrian routes and practices as a means of reading and engaging with urban space. These patrols on foot are materially and socially co-produced with walking emerging as a significant contributor to the group's instant availability.

A significant feature of these routes and activities is their repetition that aligns with the rhythmicity of urban outreach work detailed by Hall (2010). The extract highlights how these patrols show an awareness of the rhythms of the night-time economy and how Pastors, through regular patrols, develop a good knowledge of the social geographies of their towns and cities. Thus, early in the evening they might take the opportunity to talk to the staff of take-away outlets or door staff while they were not busy. Later, Street Pastors position themselves at places that are busy, such as pubs at closing time or taxi-ranks, in order to maximise their opportunities for encounters. For as one Street Pastor explained, 'the rationale for that is that that's where there are lots of concentrations of drinkers and partygoers and revelers'. The seeking out of these encounters can also be understood in relation to concerns with the right to mobility due to the overt strategy Street Pastors adopt in their search for meaningful encounters, predicated on having the right to move through urban space. This right to move through urban space has developed through, and is also predicated on, their relationships with other actors in the night-time economy, which has resulted in them having privileged access to places and opportunities for encounters via their pedestrian patrolling mobilities.

In contrast, the patrolling routes of Sock Mob volunteers, while planned, emerged in relation to a different set of concerns. The below extract is from an in-depth interview with one of the founding members of Sock Mob who is responding to a question about how the routes of their patrols developed:

JM: And you started, you started with a walk around Kings College. Initially it was the Strand, around the Strand or –?

ANNA: No, actually. It started off round the Tottenham Court Road area. Oxford Street and Euston. Because that's where a couple of these – my friends from the other church, they were kind of around there, in Bloomsbury. And then over time, yes, we came down to the Strand and Waterloo, just as we explored the city. I think that's another reason why I love doing it; that's part of the curiosity. It wasn't just the homeless; it was seeing the city in a different way. I love just walking around exploring. And I thought, "Oh, this will be an interesting way to see things." And of course I did the Cities Masters at Kings. I mean I've always been interested in that. And I think this is a very practical non-theoretical way I can really experience the city in a way that no one else is. So there was a lot of personal going on there. You know.

Anna explains the routes and timing of Sock Mob patrols by outlining where they used to walk and tracing this through to where they patrol now. She makes the link between the expansion of the walking route to their growing explorations of the city before expanding further as to why exploratory urban walking is for her 'another reason' why she 'love[s]' doing Mobbings beyond simply a motivation to engage with 'the homeless'. Anna frames the Mobbings on foot as a way of 'seeing the city in a different way'. A series of justifications follow as to the conscious adoption of such an approach informing her pedestrian patrolling activities. These include emphasising her longstanding interest in cities, evidenced in mentioning her masters studies, while framing urban exploration on foot as unique method to 'experience the city in a way that no one else is'. Anna's account can be situated in the context of a long tradition of work that draws upon walking as a method for engaging with the city (see Chapter 6 for a more detailed discussion of walking methods). For example, Back (2017) invites us to consider urban walking as a method for understanding postcolonial London in the footsteps of those protesting about 'competing visions of postcolonial London life' (21). In doing so he argues that 'walking is not just a technique for uncovering the mysteries of the city but also a form of pedagogy or a way to learn and think not just individually but also collectively' (21).

Positioning pedestrian patrolling activities as a method of knowing/ engaging with urban space also emerged in Street Pastor accounts. In the below interview extract, Emma is asked to reflect upon how her engagement with the city has changed since becoming a Street Pastor:

JM: How do you think, it's a year and a half you've been doing this, how do you think it's changed your sort of, if you like, not perception but your engagement with the city, how you see the city and as you move through it, do you notice anything?

EMMA: Well one thing that's clearly different is actually, that I see the city much more. I generally drive around the place so I don't see what is there as in sort of where different groups of houses might be, or where, you know walking around the streets at 4 o'clock in the morning is fantastic, I mean it really is lovely walking around somewhere and seeing what it really looks like and I see buildings that are different every time I go.

Emma begins by contrasting her experiences on foot when on Street Pastor patrols to her everyday journeys by car. She makes relevant the different temporalities emerging from walking as opposed to driving (see Venturi et al., 1977), which can be understood as highlighting the dominant car cultures in many cities. However, there is more at stake in Emma's account than promoting the virtues of walking over driving as a mode of urban transport. As Emma enthuses about how 'fantastic' and 'lovely' it is to walk 'around the streets at 4 o'clock in the morning' and see 'what it really looks like', it is important to consider a broader context where, for many women, to walk at night remains incredibly prohibitive (see Brands et al., 2015; Koskela and Pain, 2000). For as Elkin (2016) suggests, 'Today, when most women you meet in the city have a tale or two of street harassment to tell, the notion of wandering the streets alone seems a fraught proposition'.

As Anna and Emma both extoll the pleasure they experience in the forms of urban exploration emerging from their patrolling activities, it is important to hold this backdrop in tension with their accounts while acknowledging the right to mobility they are able to exercise. For the exposure involved in 'seeing what [the city] really looks like' cannot be divorced from the vulnerabilities experienced by different walking bodies I drew attention to in the previous chapter on the rhythms of walking. One could argue that it is the collectiveness of the 'teams' they walk with and the purposefulness of their activities which limit their exposure to such vulnerabilities. While Emma positions 'houses' and 'buildings' as particularly significant material encounters in her nocturnal walking activities as a Street Pastor, in what follows I highlight the significance of other material objects to the encounters unfolding through the pedestrian patrols of both groups.

Performing encounters: tokens and management

Members of both groups are motivated by an overarching aim to use pedestrian patrols as a method of seeking out encounters:

> On this inaugural patrol, the Street Pastors were sent off to opposing corners of the square with instructions that each person/group has to speak to at least three individuals. In other words, each person has to have three 'significant' encounters. They are instructed not to evangelise but to give 'good information' and not initiate conversations about God or religion 'that's unless they (the person encountered) wants to carry on the conversation'. As we walk around observing the different groups on this first patrol, an observer from the Ascension Trust raises some key issues which he feels need working on in future patrols. He points out how they were walking too quickly and the people they do approach are 'easy targets'. Both of these issues mean they were 'missing people, potentially more vulnerable people' and missing opportunities for 'successful interactions'. He also explains how they are sticking together as a group too much and not breaking off into pairs in order to cover more ground.
>
> (JM observation, large city)

As the above field diary extract illustrates, the importance of encounters to Street Pastors operation is such that each 'significant' social interaction occurring on a patrol is logged using record cards noting the nature of the exchange and is usually supported by supplication from Prayer Pastors. 'Significant' or 'successful' interactions are those that involve a conversation rather than just a greeting. The mobile nature of patrols is a key feature of Street Pastor interactions. Routes are planned to allow Pastors to patrol at a slow walking pace, giving them time to both meet and engage with people. This mobile dimension of their activities also enables them to approach and retreat from interactions while maximising their opportunities for a greater number of encounters in comparison to operating from a fixed station.

At this point, it is important to note that all the Street Pastors we came in contact with through the course of the research were strongly motivated by a strong ethic of care to help those who may need assistance during the course of a night out. Furthermore, on the whole, the other agencies we spoke to (the police, door staff, licensees), after acknowledging their initial scepticism, were very positive about the extra support Street Pastors have provided in city centres on a Friday/Saturday night. However, the above extract draws attention to a politics of engagement involving a series of judgments as to not only who is 'vulnerable' but also what being 'vulnerable' actually is. Although the extract highlights a strong emphasis on verbal encounters, it also draws attention to the embodied nature of these interactions in relation to the ways in which the Street Pastors walk in terms of walking slowly and not in a large group. The bodily comportment of patrolling activities is something to which I will subsequently return. However, first, I wish to draw attention to the distribution of goods/items being used by both groups as a common method of encounter engagement.

Religion is clearly at the core of the Street Pastor's practice but is not immediately apparent in their patrolling encounters. Rather, a strategy of encounters, through their pedestrian patrols, is centred around a politics of engagement which provides opportunities for such discussions to emerge and unfold, in particular, through the distribution of certain objects. The significance of these objects is how they are co-opted into becoming part of a right to encounter. These are not neutral objects but are central features in the performance of this right to encounter. For example, verbal interactions are enabled and aided by the 'kit' carried by Street Pastors. A blue, corporate uniform is the most obvious example of this. The warm, waterproof, practical clothing contrasts noticeably with the ways in which clubbers are dressed, allowing Pastors to be identified from them and, in turn, allowing interaction. During one patrol we observed, one Street Pastor was very concerned that they had misplaced their Street Pastor baseball cap. As well as being a significant part of their uniform received upon the successful completion of training, they also make Pastors easily identifiable to CCTV operators and police. Upon being probed further, the Street Pastor also explained that revellers frequently 'stole' the hat to try it on. Without it, the potential resource for interaction was lost.

Indeed, a strategy of giving away objects proved a remarkably successful way of enabling encounters. Water is given to dehydrated drinkers, space blankets to the cold, and flip-flops to women who find it difficult to walk in high heels as an evening progresses. For some, receiving something from a Street Pastor is a trophy of a 'good night out' and we witnessed countless requests of 'have you got any flip-flops?' from both women and men. It was later revealed that the volume of men asking for flip-flops (usually as a joke) had become so great that some Street Pastor groups had started handing out lollypops in response to these requests. The rationale was that although some people did not actually need alternative footwear, it was important that they were given something as to turn people away empty-handed might be perceived as rejection. Although these gifts are of immense practical value, they also provide a reason for people to talk to Pastors, for as Laurier et al. (2006) highlight, 'under most circumstances city dwellers do not initiate conversations with people with whom they are unacquainted unless by way of some legitimate mechanism that provides basis for a conversation' (14). Objects such as flip-flops provide a legitimate resource for conversation, or what Sacks (1992) refers to as a 'ticket'. 'Tickets' such as flip-flops are a mechanism for a conversation and also open up the possibility of conversations about faith. It is important to note that Pastors are explicitly forbidden by their teams to preach or attempt to convert the public to their faith, a regulation that is strictly enforced by team leaders. Pastors are, though, encouraged to articulate that they are a Christian organisation. So, when people inevitably asked Pastors what they were doing, who they were, and why they were giving away goods, the standard response was that they were from 'local churches and helping to keep people safe'.

In many ways, the use of tokens/gifts by Sock Mobbers as an icebreaker echoes the strategies deployed by Street Pastors. The below is an extract from my field diary notes from a Mobbing I participated in during the research:

> Having given out OJ and a pair of socks. I'm beginning to regret not bringing out flapjack or cakes – they stretch a lot further and I'm beginning to realise that it won't be long before I run out of items to hand out. I am surprised that Adrian (a fellow Mobber) doesn't seem to have brought anything with him. Although Sock Mob is about conversation and dialogue as opposed to distribution of resources, I begin to feel uncomfortable with this – especially as his level of engagement/conversation isn't that forthcoming either.
>
> (JM field diary entry)

The above account is framed around my anxiety about not having enough items to distribute on the patrol and in doing so reflects the central role that objects have acquired in the pedestrian practices of Mobbers. This centrality of objects to the homeless encounters Mobbers seek on foot is further re-enforced in my expressions of surprise and unease about another Mobber on that patrol having nothing with him to distribute. I attend to this criticism by pointing out that it is not just that he has nothing to give to the homeless people we encounter on our route but also that he is not actively participating in the interactions that are central to the Sock Mob's patrolling activities. This absence of interaction leads us to the question of non-encounter, the right to resist encounters, and the moral codes relating to appropriate conduct and ideas of who or what belongs in these spaces patrolled on foot. In similar terms to the Street Pastors, the items that Sock Mobbers bring with them act as a bridge to conversation. These objects do all sorts of interesting 'work' in relation to how Sock Mob encounters are practically accomplished. For example, objects keep encounters alive from week to week as requests for things such as sleeping bags or clothes are made. The objects handed out also get re-read through the street in scenarios such as water bottles being used to store alcohol. However, it became clear through our observations that these material objects were so much more than icebreakers. As such, for some Mobbers, the items they hand out make them feel they have a right to encounter and in many cases they rely heavily on the items to drive the conversations that are fundamental to their encounters with homeless people on the street.

Resisting encounters: the politics of non-encounter

There are tacit rules of engagement for the encounters that emerge from the pedestrian patrolling activities of both Street Pastors and Sock Mobbers. For example, women tend to be approached first and foremost by female Mobbers. Furthermore, visible signs of difference and encounters with difference are key to these approaches whereby there is a constant negotiation of visual stigma, which often includes situations of misrecognition when someone is judged either correctly or incorrectly as homeless:

> We approach Connexions and Samuel says he's just going over to the couple on the steps. I quickly say that I don't actually think they are homeless – which I'm sure they aren't. He takes a second look and agrees with me. We then walk past Dora – a young woman – who is reading what looks like a journal paper. We ask her if she's ok and her reaction is so quick that she is fine – almost seems affronted that we've assumed she is homeless – that we apologise and quickly move on. When we stop at the next guy waiting for Connexions to open she then comes over and clearly wants to see what we are giving out. Both instances made me think about the judgments we make about whether someone needs assistance or not – and what it is that makes someone appear in a certain way.
>
> (JM field diary entry)

The above extract from my field diary illustrates how it is visible signs of difference that go some way in explaining a selectivity as to who is, and who is not, approached during Sock Mobbing patrols. This may be related to visible items such as sleeping bags or a paper cup collecting coins from passers-by. Yet, more often, the cues are more subtle, such as where and how someone is sitting. However, there are also clearly regulars and favourites on each different route who are frequently sought out by Mobbers. This raises some fundamental questions around the recognition of this category of encounter in relation to who is recognised as deserving, or not, of contact and thereby given preferential treatment. The above extract also highlights how the form of a Sock Mobber's approach and its opening moments abide by the norms of civil engagement including smiles, nods, and the sharing of information such as names. Yet this not only extends to what is said but also to the Mobbers' embodied practices. Many Mobbers seem to be aware that the ways in which their bodies move are significant features of their homeless encounters and will make a difference to how these encounters unfold. For example, the field diary extract below illustrates how sitting down as opposed to standing is frequently used as a strategy for breaking down potential barriers:

> Adrian stands back and appears to tower over the people sat down. In the short time I've been doing this, I've begun to realise the importance of how you approach people and how you position your body as you're talking. In my view Colin is great at this – and always crouches down. From very early on, I've found myself mirroring this and am finding it a very simple and effective way of acknowledging some of the inevitable uneven power dynamics at play.
>
> (JM field diary entry)

For Mobbers, such as Colin, their bodily performance can be argued to resonate with Ignatieff's (2001) point that, respect and dignity are a matter of human gestures, not of abstract values. Concerns associated with the bodily performance of homeless encounters are well documented (see for example Hall and Smith, 2013, on homeless urban outreach work and Jackson, 2015, on the everyday mobilities of young homeless people). However, what is interesting in the case of the Sock Mob is the variation of embodied practices, perhaps due to a lack of formal training, whereby some Mobbers will stand back and almost seem shy of engaging with the 'homeless other', while others will crouch down and sit with homeless people and perhaps share a cup of tea. These encounters, and non-encounters, are disruptions to the rhythms of the Sock Mob's pedestrian patrols that thread through the swarms of people perhaps heading home on their daily commute or about to start an evening out. They are significant to understandings of urban walking through illustrating not only how walking is socially and materially co-produced but also how the rhythms and temporalities of different pedestrian practices co-exist in the city (see also Chapter 4). Yet, they also highlight how urban walking is not an innocent practice but a form of everyday mobility that has its own politics intrinsically linked to the both the actual, and potential, encounters that unfold on foot and who has the right to perform or resist these encounters. The power dynamics at play in the bodily performances of pedestrian encounters is key here and, more broadly, essential for understanding the socially differentiated nature of urban walking.

For Street Pastors, the significance of the embodied nature of their pedestrian encounters takes on a spiritual dimension. As practising Christians, many articulated how their beliefs had led them to places where they considered people were in need. For example, Simon recounts arriving at a park bench the moment a blanket blew away from a sleeping homeless man:

> And as I open the gate to go in, make sure he's okay, the space blanket lifts off with the wind and falls on the floor. I went over to him and the

best way of describing him is he had a drunken snore. He was laid out on the bench. And my water bottle is at his head. And I picked up the space blanket and I folded it round him and I tucked it in, all the way along, and I had to walk away. If I hadn't turned up at that second, see I do believe in God-incidences, not coincidences, we turned up at the very second the space blanket blew away.

(Simon, Street Pastor group leader, large city)

For this Street Pastor, his actions were not only guided by his training and equipment but also positioned through articulations of faith and his belief in 'God-incidences'. It illustrates the importance placed by Street Pastors on understanding urban-nightscapes as spiritual and, at the same time, material places. Encounters are understood in the same way and are often positioned as having religious significance. As Dewsbury and Cloke (2009) note, this reflects a 'spiritual landscape' in which experiences are understood in relation to 'bodily existence, felt practice and faith in things that are immanent' (696).

However, more is at stake in this account. In describing his actions, Simon recounts the active work he did in this unfolding act of care: 'I open ... I went ... I picked up ... I folded ... I tucked...'. The man described as having 'a drunken snore' is a passive recipient of these acts of care, or what Simon constructs as godly acts of divine significance, as he continues to sleep. As Simon presents his account of a social encounter he experienced while patrolling on foot, what in fact emerges is a clear uneven distribution of power. The park bench and its occupant were accessible by foot. It was the very nature of walking that enabled Simon, and gave him the power, to approach. This example not only raises questions as to how care is offered, distributed, and received in these contexts but also highlights the subtle power relations of urban walking in public space and the politics of such (non)encounters. Who has the power to initiate or avoid such encounters? Who has the privilege to resist encounters on foot and what form does resistance to pedestrian encounters take? Such an example further emphasises how walking is socially differentiated with inequalities emerging from people's diverse experiences of everyday urban walking.

To understand the politics of encounters and non-encounters of everyday walking practices, it is useful to further consider a series of associated moral codes. In the examples of Street Pastors and the Sock Mob being discussed here, it relates to appropriate recipients of specific forms of care but it is possible to extend this point to the broader context of how people appropriate space on foot and the moralities that emerge from and inform these practices. For example, the extract below is from my field notes taken during the inaugural Street Pastor patrol in one large UK city:

Just after 10pm – walk to ***** area of city. Andy explains that this is where the police wanted them to patrol initially. The SPs originally wanted to patrol the "gay district" but this was deemed too "risky" at present due to "churches relationship/attitude to homosexuality". Particularly since one of the local Catholic church leaders had spoken openly about his opposition to same sex relationships. So it was decided that they would start in ***** and then eventually spread out into other areas as more people were recruited. Frequency of patrols at present will be every other Friday night.

(JM observation, large UK city)

This extract highlights not only the pre-planning that informs where Street Pastors patrol – these are not serendipitous walking routes – but also the significance of the types of encounters that might unfold. In this case, the route is planned in order to mitigate the likelihood of what have been considered as potentially 'risky' encounters between the Christian volunteers and LGBTQ community members. There were genuine concerns that the Street Pastors would encounter a hostile reception to their presence in areas of the city where gay bars and nightclubs were located due to the enduring prevalence of conservative attitudes to same sex relationships within the church. There are clear parallels here with forms of non-encounter experienced in the context of enclave urbanism (see Schuermans, 2016) in relation to a deliberate spatial segregation of different groups. Through focusing on pedestrian routes such as these and the practices of Street Pastors a particular politics of engagement becomes evident with both encounters and non-encounters contributing to a series of unequal outcomes whereby the subtle power relations of walking through urban public space are rendered more visible. More specifically, a politics is apparent as to who has the privilege of performing and resisting encounters.

Conclusion

Urban walking is a social practice insofar as it involves walking with, past, in tandem, against, and through other people. Walking encounters are part and parcel of the socio-material co-production of everyday walking practices, yet pedestrian non-encounters are also significant for understanding the everyday politics of how people appropriate urban space. Although non-encounters are more difficult to examine empirically and, as a result, often overlooked in favour of what is visibly encountered, the encounters which are avoided, missed, absent, or resisted are an important further dimension of urban walking that prompt us to consider differently pedestrian socialities. This chapter has engaged with Kärrholm et al.'s (2017) invitation to reflect on how different sorts of walking practices are assembled to co-exist (or not) in order to gain new insights into the subtle power relations of urban mobility in public space.

Street Pastor patrols operate in the public spaces of the night-time economy and can be called by the police, CCTV operators, or other actors to deal with incidences with which they require assistance. Street Pastors frequently emphasise how they have become valued by these agencies for their ability to deal with people who are vulnerable and to provide practical help, emotional support, or counselling. Examples might include walking with a lone woman who has lost her friends, administering minor first aid, or assisting a person incapacitated by drink to a taxi. It is not in the remit of the emergency services to deal with these people as they have not committed a crime (in the case of the police) or are not seriously injured thus meriting the attention of the paramedics (Jayne et al., 2010). Street Pastors' walking practices have become an extension of traditional surveillance through CCTV and the police. They function as a further set of 'eyes on the streets' (Jacobs, 1972). For as one pastor explained, 'we let everyone know we're keeping a friendly eye on them'. The encounters which unfold, or not, through these pedestrian surveillance practices are at the centre of what constitutes a 'night out' for street pastors.

For London's Sock Mob the central objective of their pedestrian patrols is to seek out encounters with homeless people. Connecting two parties who would otherwise seem to inhabit entirely different worlds, might be considered what Valentine (2008) would term a 'meaningful encounter'. As such, the most obvious and immediate outcome of such encounters is the breakdown of some of the isolation and loneliness that is a significant part of the street homeless experience while challenging the stigma street homeless people suffer. For Amin (2006) one of the difficulties in looking to the streets as a locus of interaction in the 'good city' is that streets are primarily spaces of transit, and he suggests that meaningful interaction is only likely to arise in those cases where people's movements follow well-worn rhythms and patterns that bring them in to regular contact with particular people.

In examining the pedestrian patrolling activities of both these groups, I have extended concerns with the right to the city and the right to mobility by examining the right to encounters. Who has the right to seek encounters or resist them? In doing so, I have emphasised how both encounters, and non-encounters, on foot matter to our understandings of the inclusions and exclusions that can emerge from different types of urban walking. A micropolitics relating to a series of moral codes as to who and who is not approached during these walks leads to a series of unequal outcomes as to how care is distributed and received. In particular, this focus has shown the power relations that can unfold from how people appropriate urban space on foot while challenging the ways in which pedestrian social encounters are frequently either romanticised and underpinned by a series of positive assumptions or simply considered a benignly neutral aspect of contemporary urban life. This is an important point in encouraging a closer consideration of how walking is not always a positive urban practice but can also be something to endure (see Davidson, 2020).

Pedestrian movement is central to the activities of Street Pastors and the SockMob. Yet, this is not simply a case of 'walking-as-transport' as a means of getting from A to B in order to access opportunities such as employment and leisure. Rather, walking practices are fundamental to the encounters that are central to the groups' objectives and their very existence. It is the nature of being on foot that facilitates the social interactions they seek through their patrolling activities. However, these patrolling activities demonstrate how walking is not an innocent practice but one from which emerges complex and subtle power relations. As such, we should resist further romanticising how urban walking is positioned and promoted in relation to understandings of contemporary urban life. Attending to the complexity of pedestrian encounters and non-encounters also renders visible their embodied nature. It is concerns with embodiment and the walking body to which I turn in the next chapter.

Notes

1. A person who approaches people in the street asking for subscription donations for a particular charity.
2. See https://www.streetpastors.org.
3. See https://www.meetup.com/thesockmob/.
4. A national umbrella organisation that promotes and regulates the work of Street Pastors. Local Street Pastor groups are launched at a church commissioning service that is attended by Street Pastors, representatives from the Ascension Trust, local churches, and representatives from local services, including the police and local council, emphasising their enrolment into formal city networks.
5. Funded by the British Academy.
6. However, even if being a non-Christian had not prevented me from joining the Street Pastors, I would have been uneasy about being involved with a group whose activities were so strongly rooted in religious practices such as prayer.
7. This undoubtedly raises issues around the blurring of identities as both a group member and a researcher. Developing a more formalised research project with Mobbers' consent was one way in which we attended to this tension due to it making these multiple identities more visible.

References

Amin, A. 2006. The good city. *Urban Studies, 43,* 1009–1023.

Attoh, K. A. 2012. The transportation disadvantaged and the right to the city in Syracuse, New York. *Geographical Bulletin,* 53.

Back, L. 2017. Marchers and steppers: Memory, city life and walking. In: Bates, C. & Rhys-Taylor, A. (eds.) *Walking through social research,* New York, Routledge, 21–37.

Baiocchi, G. 2001. Brazilian cities in the nineties and beyond: New urban dystopias and utopias. *Socialism and Democracy, 15,* 41–61.

Beaumont, M. 2015. *Nightwalking: A nocturnal history of London, Chaucer to Dickens,* London, Verso.

Brands, J., Schwanen, T. & van Aalst, I. 2015. Fear of crime and affective ambiguities in the night-time economy. *Urban Studies, 52,* 439–455.

Bulut, E. 2018. Race or class? Testing spatial assimilation theory for minorities in Los Angeles. *Polish Sociological Review, 201,* 127–141.

Castañeda, P. 2020. From the right to mobility to the right to the mobile city: Playfulness and mobilities in Bogotá's cycling activism. *Antipode, 52*, 58–77.

Cloke, P. J., May, J. & Johnsen, S. 2010. *Swept up lives?: Re-envisioning the homeless city*, Oxford, Wiley-Blackwell.

Cresswell, T. 2009. The prosthetic citizen: New geographies of citizenship. *Political Power and Social Theory*, 259–273.

Davidson, A. C. 2020. Radical mobilities. *Progress in Human Geography*, 1–24.

De Certeau, M. 1984. *The practice of everyday life*, Berkeley, London, University of California Press.

De Souza, M. L. 2010. Which right to which city? In defence of political-strategic clarity. *Interface, 2*, 315–333.

Darling, J. & Wilson, H. F. (eds.). 2016. *Encountering the city: Urban encounters from Accra to New York*, London, Routledge.

Dewsbury, J. D. & Cloke, P. 2009. Spiritual landscapes: Existence, performance and immanence. *Social & Cultural Geography: Themed issue: Geography, Religion, and Emerging Paradigms: Problematizing the Dialogue, 10*, 695–711.

Duff, C. 2017. The affective right to the city. *Transactions, 42*, 516–529.

Dunn, N. 2016. *Dark matters: A manifesto for the nocturnal city*, Winchester, Zero Books.

Edensor, T. 2015. The gloomy city: Rethinking the relationship between light and dark. *Urban Studies, 52*, 422–438.

Elkin, L. 2016. A tribute to female Flaneurs: The women who reclaimed our city streets. *Guardian*, 29th July.

Fenster, T. 2005. The right to the gendered city: Different formations of belonging in everyday life. *Journal of Gender Studies, 14*, 217–231.

Goffman, E. 1963. *Behavior in public places: Notes on the social organization of gatherings*, New York, Free Press.

Golub, A. & Martens, K. 2014. Using principles of justice to assess the modal equity of regional transportation plans. *Journal of Transport Geography, 41*, 10–20.

Hall, T. 2010. Urban outreach and the polyrhythmic city. In: Edensor, T. (ed.) *Geographies of rhythm: Nature, place, mobilities and bodies*, Farnham, Ashgate, 59–71.

Hall, T. & Smith, R. J. 2013. Stop and go: A field study of pedestrian practice, immobility and urban outreach work. *Mobilities, 8*, 272–292.

_____. 2014. Knowing the city: Maps, mobility and urban outreach work. *Qualitative Research, 14*, 294–310.

Harvey, D. 2008. The right to the city. *New Left Review, 53*(September–October), 23–40.

Hubbard, P. 2004. Cleansing the metropolis: Sex work and the politics of zero tolerance. *Urban Studies, 41*, 1687–1702.

Ignatieff, M. 2001. *Human rights as politics and idolatry*, Toronto, Anansi Press Ltd.

Ingold, T. & Vergunst, J. L. 2008. *Ways of walking: Ethnography and practice on foot*, Farnham, Ashgate.

Jackson, E. 2015. *Young homeless people and urban space: Fixed in mobility*, New York, Routledge.

Jacobs, J. 1972. *The death and life of great American cities*, Harmondsworth, Penguin.

Jayne, M., Valentine, G. & Holloway, S. L. 2010. Emotional, embodied and affective geographies of alcohol, drinking and drunkenness. *Transactions of the Institute of British Geographers, 35*, 540–554.

Kärrholm, M., Johansson, M., Lindelöw, D. & Ferreira, I. A. 2017. Interseriality and different sorts of walking: Suggestions for a relational approach to urban walking. *Mobilities*, *12*, 20–35.

Kern, L. 2020. *Feminist City: Claiming space in a man-made world*, London, Verso.

Koskela, H. & Pain, R. 2000. Revisiting fear and place: Women's fear of attack and the built environment. *Geoforum*, *31*, 269–280.

Kusenbach, M. 2003. Street phenomenology: The go-along as ethnographic research tool. *Ethnography*, *4*, 455–485.

Laurier, E., Maze, R. & Lundin, J. 2006. Putting the dog back in the park: animal and human mind-in-action. *Mind, Culture and Activity*, *13*, 2–24.

Lefebvre, H. 1970. *La Révolution urbaine*, Paris, Editions Gallimard.

_____. 1991. *The production of space*, Oxford, Blackwell.

Low, S. M. 2001. The edge and the center: Gated communities and the discourse of urban fear. *American Anthropologist*, *103*, 45–58.

Lucas, K. & Jones, P. 2012. Social impacts and equity issues in transport: An introduction. *Journal of Transport Geography*, *21*, 1–3.

May, J. & Cloke, P. 2014. Modes of attentiveness: Reading for difference in geographies of homelessness. *Antipode*, *46*, 894–920.

McCann, E. J. 1999. Race, protest, and public space: Contextualizing Lefebvre in the US city. *Antipode*, *31*, 163–184.

Merrifield, A. 2011. The right to the city and beyond: Notes on a Lefebvrian reconceptualization. *City*, *15*, 473–481.

Middleton, J. 2018. The socialities of everyday urban walking and the 'right to the city'. *Urban Studies*, *55*, 296–315.

Mitchell, D. 2003. *The right to the city: Social justice and the fight for public space*, New York, Guilford Press.

Mohan, J. 2000. Geographies of welfare and social exclusion. *Progress in Human Geography*, *24*, 291–300.

Morgan, L. J. 2013. Gated communities: Institutionalizing social stratification. *The Geographical Bulletin*, *54*, 24.

Paasche, T. F., Yarwood, R. & Sidaway, J. D. 2014. Territorial tactics: The socio-spatial significance of private policing strategies in Cape Town. *Urban Studies*, *51*, 1559–1575.

Plyushteva, A. & Boussauw, K. 2020. Does night-time public transport contribute to inclusive night mobility? Exploring Sofia's night bus network from a gender perspective. *Transport Policy*, 87.

Purcell, M. 2003. Citizenship and the right to the global city: Reimagining the capitalist world order. *International Journal of Urban and Regional Research*, *27*, 564–590.

_____. 2013. Possible worlds: Henri Lefebvre and the right to the city. *Journal of Urban Affairs*, *36*, 141–154.

Sacks, H. 1992. (edited by G. Jefferson). *Lectures on conversation*, Oxford, Blackwell.

Schuermans, N. 2016. Enclave urbanism as telescopic urbanism? Encounters of middle class whites in Cape Town. *Cities*, *59*, 183–192.

Shaw, R. 2018. *The nocturnal city*, London/Abingdon, Routledge.

Sheller, M. 2018. *Mobility justice: The politics of movement in an age of extremes*, London, Verso.

Simmel, G. 1971. *The metropolis and mental life*, Chicago, University of Chicago Press.

Smeds, E., Robin, E. & McArthur, J. 2020. Night-time mobilities and (in)justice in London: Constructing mobile subjects and the politics of difference in policymaking. *Journal of Transport Geography*, 82.

Smith, N. 1996. *The new urban frontier: Gentrification and the revanchist city*, London, Routledge.

Sock Mob. 2018. The Sock Mob homeless volunteer meetup group [available at: https://www.meetup.com/thesockmob/about/].

Spinney, J. 2015. Close encounters? Mobile methods, (post)phenomenology and affect. *Cultural Geographies*, *22*, 231–246.

Sukaryavichute, E. & Prytherch, D. L. 2018. Transit planning, access, and justice: Evolving visions of bus rapid transit and the Chicago street. *Journal of Transport Geography*, *69*, 58–72.

Thrift, N. 2005. But malice aforethought: Cities and the natural history of hatred. *Transactions of the Institute of British Geographers*, *30*, 133–150.

Valentine, G. 2008. Living with difference: Reflections on geographies of encounter. *Progress in Human Geography*, *32*, 323–337.

Van Liempt, I., Van Aalst, I. & Schwanen, T. 2015. Introduction: Geographies of the urban night. *Urban Studies*, *52*, 407–421.

Venturi, R., Scott, D., Robert, I. V., Brown, D. S., Izenour, S. & Steven, R. V. D. S. B. 1977. *Learning from Las Vegas: The forgotten symbolism of architectural form*, Cambridge, MA, MIT Press.

Verlinghieri, E. & Schwanen, T. 2020. Transport and mobility justice: Evolving discussions. *Journal of Transport Geography*, 87.

Verlinghieri, E. & Venturini, F. 2018. Exploring the right to mobility through the 2013 mobilizations in Rio de Janeiro. *Journal of Transport Geography*, *67*, 126–136.

Wacquant, L. 1999. Urban marginality in the coming millennium. *Urban Studies*, *36*, 1639–1647.

Wilkinson, S. 2017. Drinking in the dark: Shedding light on young people's alcohol consumption experiences. *Social & Cultural Geography*, *18*, 739–757.

Wilson, H. F. 2017. On geography and encounter: Bodies, borders, and difference. *Progress in Human Geography*, *41*, 451–471.

Wilson, W. J. 1997. *When work disappears: The world of the new urban poor*, New York, Random House.

Yarwood, R. 2007. The geographies of policing. *Progress in Human Geography*, *31*, 447–465.

6 Walking bodies, emotional geographies, and mobile methods

The Mission District in San Francisco provides the backdrop for a walk Judith Butler took with the disability activist Sunuara Taylor as part of a wider documentary entitled 'Examined Lives', which consisted of eight short films of philosophers speaking about their work.[1] The opening shots of the film feature a range of everyday mobile practices including cycling, pushing a pram, skate boarding, and walking with a stick. The camera then moves to the front of Butler and Taylor and documents their walk together through the district. This is a fascinating and powerful piece of film where the dialogue between Butler and Taylor as they move together draws attention to a series of concerns relating to gender, disability, identity, and mobility. Butler opens by posing the question of 'what it means for us to take a walk together'. This seemingly simple question provides the entry point for exploring the complexity of the identities and relations emerging from our everyday mobilities. However, it is Butler's subsequent comments that provide the opening for what I wish to examine in this chapter. First, Butler's posing of the question, 'what can a body do?', brings concerns with the body to centre stage. Second, her statement that 'nobody goes for a walk without having something to support that walk' assumes *interdependent* mobility, as opposed to *independent* mobility, as a predetermined given and the starting point for a particular politics to emerge.

This chapter examines the embodied dimensions of pedestrian practices in the context of these two concerns in relation to what supports a walk. In doing so, I consider how these relate to broader concerns with how walkability is understood and conceptualised. This book has already drawn out the significance of the lived experiences of walking through interrogating pedestrian infrastructures, the socially differentiated nature of moving on foot, temporal politics, and urban encounters. In this chapter I critically examine the embodied, material, and technological relationality of walking to further disrupt self-evident notions of walkability. I start by asking what embodiment means in the context of everyday walking practices. Of what significance are the bodily senses to everyday experiences of walking in terms of how urban space is experienced and produced? The embodied dimensions of walking are central to performative and artistic engagements with pedestrian practices. However, underpinning much of this work are privileged and universalising tendencies of walking, which fail to take account of differentiated mobile bodies. By considering the emotional and affective geographies of pedestrian practices, I attend to a neglected yet important dimension for further understanding the embodied dimensions of everyday urban walking. I move on to critical reflections on research using mobile methods, including walking methods, to examine the everyday mobility experiences of visually impaired (VI) young people in London. This group's experiences of navigating the city, while engaging with these methods, both counter and challenge the walking methods many artists and philosophers have the privilege of incorporating in their work. I conclude by discussing the opportunities and limitations of walking methods in relation to notions of accessibility and the 'demands of the method' (Warren, 2017: 802) to differentiated mobile bodies.

The neglected walking body

In 2003 I began my doctoral research on walking in the city. My research began as an Economic and Social Research Council (ESRC) Collaborative Award in Science and Engineering (CASE) studentship[2] with Arup Transport Planning (now known as Arup). The involvement of Arup in the research stemmed from their interest in an urban environmental modelling tool (CityMatters), designed to investigate walking in the built environment, which belonged to and was being developed by a consultant within the organisation. In the context of my research, the model was used as a tool to explore the types of urban environments that encourage walking. CityMatters was a descriptive model that evolved from Space Syntax, 'a set of theories and methodologies for analysing spatial systems developed by a research unit at the Bartlett School of Architecture, at UCL' (Alayo, 2002: 4). Alayo (ibid.) points out how 'the main hypothesis underlying Space Syntax techniques is that there is a strong relationship between the structural properties of the spaces within a system (in the sense of how they relate to all others) and the way people use them' (5). Both CityMatters and Space Syntax use the concept of 'depth' to explore the strong correlations between the distribution of pedestrian flow patterns in urban systems and the visibility characteristics of the network (Alayo, 2002). In other words, 'depth' relates to a person's line of sight in terms of the number of changes of direction necessary to reach one location from another. The software also had the built-in capability of taking account of the type and density of the land uses in each area. When lines were plotted onto a base map of a specific area that represented lines of sight along all the streets within that area, this information could be input into the CityMatters software and detailed data concerning pedestrian accessibility was produced.

As I began to familiarise myself with and learn to use the software, my unease with this approach to understanding walking in the city grew. What about other senses? How did smell, touch, and hearing relate to such an approach? What about those who do not experience the urban environment in relation to 'depth' and lines of sight? Where do differentiated bodies fit into such a model? I pursued these concerns through other approaches I incorporated into the research, namely the use of a walking photo diary/interview method (see Chapter 2 for more detail). The intellectual property rights of CityMatters belonged to one particular consultant, who was the only person within Arup who had any in-depth knowledge of the software. When he left Arup to work overseas 18 months into the studentship, the collaboration and use of the software disintegrated. At the time, I was relieved to now have the freedom to fully dedicate my efforts to exploring what I considered were more pressing concerns with pedestrians' everyday experiences of navigating the city on foot. Yet, in the subsequent years I have frequently thought about this experience in terms of what the model overlooked, and how it has shaped and motivated my lines of enquiry into walking in relation to rendering the visibility of walking bodies. In other words, how can we encourage the importance of our pedestrian bodily experiences to be taken more seriously?

The senses, embodiment, and the walking body

In the last 40 years there has been a broad shift across much of the social sciences relating to how the body is understood and theorised. The term 'the body', as a single, universal, pre-inscribed, and finished product, has increasingly been replaced by the term 'embodiment', whereby the body is understood as produced in relation to, and 'made up' by, larger networks of social and cultural relations (Cresswell, 1999; Grosz, 1994; McDowell, 1995). Much of this work on embodiment is situated in phenomenology (Ash and Simpson, 2016; Grosz, 1994; Lorimer and Lund, 2003; Wylie, 2002, 2005) and how the world reveals itself emergently in the way that it feels, rather than as fixed and static, awaiting perception. In Seamon's (1980) writings concerning a phenomenology of everyday life, he argues for an engagement with a 'bodily intentionality' called 'body-subject':

> Body-subject is the inherent capacity of the body to direct behaviours of the person intelligently, and thus function as a special kind of subject which expresses itself in a preconscious way usually described by such words as 'automatic,' 'habitual,' 'involuntary,' and 'mechanical'. (155)

For as Hubbard (2006) elaborates, 'the implication here is that we do not have to think about the way we move through urban space: our body feels its way' (119). Seamon (1980) builds upon his notion of the 'body-subject' with 'body-ballets' where 'basic movements of body-subject fuse together into wider bodily patterns that provide a particular end or need' (158) and 'time-space routines' which are 'a set of habitual bodily behaviours which extends through a considerable portion of time' (158). In a related vein, as I discussed in Chapter 4, in the writings of Lefebvre (2004) on everyday life and rhythm, a significant emphasis is placed on the role of the body in terms of how 'rhythms converge on the body in the everyday' (Horton, 2005: 73) while it 'serves as a metronome' (Elden, 2004: xii). For writers such as Simmel (1971), being unaware of the body, or the body shutting down, is a mechanism for urban dwellers/pedestrians to cope with the potential sensory overload of the city. It is therefore possible for urban spaces to be inhabited without necessarily being sensed, as 'sensation, sensual presence, is still more, more than embodiment' (Feld 2005: 181). There is thus an analytic distinction to be made between the embodied and the sensory.

There has been much written in recent years concerning the body and senses and how these can shape our urban experiences (Adams and Guy, 2007; Borer and Borer, 2013; Degen, 2014; Low and Kalekin-Fishman, 2018; Pow, 2017). These concerns with how the sensual experience of everyday life is sensorially mediated through sound, smell, touch, taste, and sight are reflected in what has been described as the field of 'sensory urbanism' (Jaffe et al., 2020). An early example of such work drawing upon a multi-sensory approach is Adams et al.'s (2007) examination of residents' sensorial experiences of the 24-hour city in the city centres of Manchester, Sheffield, and the Clerkenwell area of London. They argue that 'the 24-hour city is a place rich with sensorial encounters and that these are highly significant components of people's everyday urban experience' (201). The problematic nature of privileging one sense over another and that embodied experiences need to be understood as multi-sensory is widely recognised (Crary, 1999; Degen et al., 2008; Ingold, 2004; Low, 2012; Pallasmaa, 2005; Rodaway, 1994; Saldanha, 2002; Vannini et al., 2013). In particular, this work questions the dominance of the visual register over other senses such as touch and smell. For, as Ingold (2000) posits, 'looking, listening and touching, they are not separate activities, they are just different facets of the same activity: that of the whole organism in its environment' (261) (see also Macpherson, 2006, on the ocular-centrism of much work on rural landscapes).

The growing literature across the social sciences and humanities on embodiment and walking considers the ways in which pedestrian practices can be adopted to create 'new embodied knowledges of place' (Doughty, 2013: 141), while centring embodied encounters with place. For example, Pink et al. (2010) draw together the significance of walking, artistic practices, and ethnography to such concerns. They illustrate how the visual needs to be 're-situated as an element of the multi-sensoriality of everyday contexts' (4) (such as Edwards and Bhaumik, 2008; Mitchell, 2002; Pink 2009), while stressing the importance of the 'visual' as 'always contextualised through the multisensory and mobility that characterise everyday experience' (4). They consider that walking as an artistic practice and as visual ethnography 'brings to the fore the interrelatedness of the visual and other senses' (see Pink, 2009: 4–5).

In Doughty's (2013) work on led walking groups in the South-East of England, she considers the potential of shared movement in producing therapeutic landscapes and how 'the bodily act of walking styles social interaction in specific ways' (142). Through an ethnographic exploration of these groups in the English countryside, Doughty describes 'bodily communicative acts' (143) in terms of how walkers negotiated social relations of the walking group. She concludes by advocating the shared practice of 'walking-with' as beneficial in the pursuit of 'wellness' (145). For Stevenson and Farrell (2018) 'walk along' interviewing was drawn upon to understand the experiences of leisure walkers on the South Downs Way, a long-distance trail in the South of England that runs for 160 km from Winchester in Hampshire to Eastbourne in East Sussex. They argue that this approach was particularly effective for engaging with the 'bodily sensations and emotional states associated with the leisure walking experience' (429). I will return to concerns with walking methods but at this point wish to continue examining the significance of walking bodies.

Parental walking bodies

The bodily act of walking was brought into sharp focus for me during pregnancy and the year following the birth of my first child. It was through this time that I became acutely aware, despite having engaged with concerns relating to walking and embodiment for several preceding years (see for example Middleton, 2010), of the differentiated mobilities emerging through our embodied experiences of moving through the city on foot. Issues with low blood pressure in the latter stages of pregnancy made walking anywhere a challenge. Following the birth I pounded the dark January streets, too scared to stop the pram moving or go somewhere warm for fear of the baby waking up and needing yet another painful feed, freezing hands trying to eat while pushing the pram, and feet and legs feeling like lead weights through sheer exhaustion with the wind blowing into tired stinging eyes. It was in these often dark first few months of becoming a mother that I began to reflect upon the challenges of negotiating urban space as a new parent. For, as Adey et al. (2012) highlight, 'To parent is to perambulate. In the early days of course it was less about mobility, and all about sleep. Or it was about how different ways of becoming mobile might facilitate becoming sleepy' (174).

The central concern of the piece of research that developed in the context of my early parenting experiences was the everyday practices of new parents in the context of urban austerity[3] (see Middleton and Samanani, 2021), yet a key theme emerging from this work relates to mobility and immobility, particularly on foot. For example, in the interview extract from the research below, Lily, one of the research participants, does all sorts of work positioning her baby as an independent being conscious of her own actions. This narration of power dynamics can be understood as a strategy for making sense of the unpredictability of small babies. However, concerns with mobility and immobility are also central to her account:

> If you stop walking, she's like ... she starts snuffling and knows that you've stopped walking and fidgeting and if you sit down she knows. I have managed to get away with it a couple of times in the day of being like able to go out for 10 minutes and then coming back here and sitting on the sofa for half an hour and she'll sleep there. But generally, in the evenings [baby noises]. Yeah [laughs]. She's a little bit more grumpy and just needs walking ... She basically does the opposite of what you want her to do.
>
> (Lily – baby, two months)

Infants both constrain and demand mobility, with many parents finding their experiences of the city transformed by having children, and discovering that certain elements of the urban form support their mobility needs, while others make these more difficult to realise – benches, footpaths, road traffic, huge crowds of pedestrians, a lack of dropped curbs, a lack of changing facilities, no elevator for the buggy, and a lack of places to keep a buggy if stopping. Over the last 20 years there has been work considering how families journey together in cars (Laurier et al., 2008) and how transporting children in cars relates to concerns with 'good mothering' (Barker, 2011; Dowling, 2000; Goodwin and Huppatz, 2010).

More recently, work has considered in closer detail the technologies involved in becoming mobile as a parent. For example, Jensen (2018) argues the pram is an overlooked design object that affords urban mobility. He draws upon an ethnographic account of his own 'pram mobilities' as a new parent in advocating a 'mobilities design approach' planning accessible urban spaces where 'the affective, the atmospheric, the social, the cultural and the material, and how these "soft" components are central parts in the creation, stabilization and reimagination of urban mobility systems' (14). Whittle (2019) explores the impact of parents using slings on everyday mobilities in the early lives of families and its potential as a technology to overcome the challenges associated with becoming mobile with infants and young children. Through Jensen and Whittle's writing, the significance of the embodied experiences of walking emerges, yet parents are positioned together with little differentiation between the experiences of mothers and fathers. Gendered experiences and the pressures which shape maternal subjectivities are clearly prominent features of Whittle's research, yet remain implicit concerns in her particular account.

In contrast, Clement and Waitt (2017) explore how specifically motherhood is felt on-the-move in car dependent cities. They are concerned with the walking body and theorise motherhood 'as part of a mobile "mother-child-walking assemblage"' (1189) (see also Boyer and Spinney, 2016). Their empirical work in the Australian city of Wollongong examines how the dilemmas of 'good mothering' are encountered and addressed on foot. In positioning walking as 'an embodiment of gendered care', they argue that the 'walkability of cities must be understood through how moving bodies experience the materiality of place (noisy/quiet streets, speeding cars, pavements, parks) in relation to the felt affects of mothering' (1199). This work highlights the significance of the body to pedestrian experiences is not just sensual but how time, space, and place are emotionally apprehended. However, despite Clement and Waite's important intervention, the ways in which urban walkers use and take account of the emotional and affective significance of their embodied experiences of place warrant further attention. In particular, how do issues of anxiety, fear, and being 'out of place' become occasioned in relation to emotion and affect?

The emotional geographies of walking

Davidson and Milligan (2004) argue emotion should be taken seriously 'since there is little we do with our bodies that we can think apart from feeling' (523). Places are related to not only through senses such as smell and touch but also through feelings, affects, and emotions mediated through these sensory engagements. Geographers have become increasingly engaged with how emotionality is central to human experience (see Anderson, 2004; Bondi and Davidson, 2016; Tolia-Kelly, 2006). The establishment of the journal *Emotion, Space and Society* in 2008 perhaps best exemplifies the institutionalisation of this interest. At the heart of the publication lie interests relating to the 'emotional intersections between people and places'. Yet, concerns with emotion and affect have been approached from a variety of different perspectives. Throughout my academic career I have been witness to numerous heated debates at conferences, in seminar rooms, and over coffee and lunch breaks in relation to some of these differences and, in particular, associations with the so-called 'non-representational turn'. For, as McCormack (2006) reminds us, affect and emotion are understood in different ways by different people, which have conceptual, political, empirical, and ethical implications.

An early example of such divergences played out across the pages of the geographical journal *Area*, with Thien (2005) voicing concerns about the 'masculinist, technocratic and distancing ways' (452) notions of affect that have been deployed at the expense of intersubjective processes and 'issues of relationality which are so profoundly embedded in our everyday emotional lives' (453). Both McCormack (2005) and Anderson and Harrison's (2006) responses argue that Thien caricatures the work she critiques without taking into account the diversity with which emotion and affect are engaged. A few years later, Steve Pile's (2010) account of 'Emotions and affect in recent human geography' in the geographical journal *Transactions of the Institute of British Geographers* also sparked debate. Bondi and Davidson (2011) critique Pile's dualistic framing of emotion and affect, while Dawney (2011) and Curti et al. (2011) share their concerns about certain conceptualisations of affect, including the work of Spinoza, being neglected. It is not my intention to rehearse, or situate myself in, these debates here but rather to use them to highlight the wide range of work on affect and emotion.

Despite such heterogeneity, much work engaging with 'affect' pays particular attention to the pre-cognitive 'how' of emotional relations (Anderson and Harrison, 2010). McCormack (2010) distinguishes between affect, feeling, and emotion in suggesting that 'affect is a kind of vague yet intense atmosphere; feeling is that atmosphere felt in a body; and emotion is that felt intensity articulated as an emotion' (1827). However, in the following analysis I demonstrate how emotion and affect is used as a discursive resource (Edwards, 1997). In particular, how emotional and affective categories are used in talk and text about other things, in this instance walking. Wetherell (2013) proposes the necessity of understanding affect and discourse as intertwined. In doing so she draws attention to the work of Goodwin (2006) on the situated activities making up playground practices to highlight how embodied action is bound up with talk in the flow of activity and that 'this entangling may occur either through patterns of utterances associated with activities in the moment or occur subsequently as participants begin to account for, communicate and make sense of their actions' (360). It is the latter 'entangling' that I wish to attend to here.

While the analysis that follows would not necessarily be recognised by discourse scholars as discourse analysis in its purest form, with all associated transcript conventions of utterances and pauses, and so on, it does draw upon some core principles. In particular, the interview data has been analysed from the perspective of understanding talk as a social activity with attention being drawn to 'members' categories' (Potter and Wetherell, 1987; Sacks, 1992). In other words, the analysis is concerned with the discursive orientation of participants' accounts and, in contrast to a top down analysis of pre-defined categories, the emergence of what they themselves take to be significant. No a priori distinctions are drawn, say, for example, between emotions, feelings, or affective reactions as what matters is what people do with talk of emotion, whether as Edwards (1997) points out in terms of 'avowing' their own or 'ascribing them to other people' (170). My aim is to examine 'how emotion discourse is an integral feature of talk about events, mental states, mind and body, personal dispositions and social relations' (ibid). Of course, it is possible to conceptualise emotions in advance. For example, Hubbard (2005) argues that 'emotions can be conceptualised as the felt and sensed reactions that arise in the midst of the (inter) corporeal exchange between self and world' (121). Such distinctions are most certainly of interest. However, the discursive turn on emotion is to examine how people actually use emotional and affective distinctions in accounting for their being in the world and, in this context, how people use and account for emotions in relation to their embodied pedestrian experiences. The emotional geographies of walking are neglected concerns across most engagements with urban walking. Furthermore, even within work on embodiment and walking, the significance of emotion and affect to embodied experience can be overlooked or sidelined as of secondary importance.

Using emotions

The extract below was drawn from in-depth research introduced in Chapter 2 on the everyday experiences of walking in London. This London Fields resident was asked 'what types of things affect your feeling of personal safety when walking?':

> I definitely feel safer when I'm not looking so feminine, you know shoes I can run in, and I'm also a fast runner so that gives me a sense of safety. I normally wear flat shoes but I do enjoy wearing heels sometimes when I go out and I really hate ... actually that's something, I had a really frustrating night out when I went out to the ICA a while back in the summer and I was wearing high heel shoes and it was Friday night and I wanted to go home. I was really tired waiting for the bus in high heels but the bus was not coming at all and I would have started walking but I couldn't walk as I had these stupid shoes on and then finally a bus came and there were drunken people everywhere all around me, students and stuff, and I was like I'm nearly 30, I'm so beyond this now, like getting really pissed off that I don't have a job or I could have got a taxi. I thought 'I'm too old and I'm just going to get a corporate job and sod the art'. So I was so annoyed, I was annoyed to tears. I finally got on this other bus and the bus driver's like 'alright love?' and you know started chatting me up and I was like so disempowered in these stupid shoes but it's so weird because you feel very empowered by them in a certain context because you feel feminine and your femininity can be empowering but then in the wrong context it totally takes everything away from you.
>
> (interview – Emma, London Fields resident, Hackney)

The issue of wearing high-heeled shoes is a significant dimension of Emma's account. In a study of female merchant bankers, McDowell (1995) investigates bodily performances in relation to the multiple ways in which aspects of both masculinity and femininity are negotiated. Emma's description of how her shoes relate to how she feels empowered in some situations and disempowered in others highlights the issues raised by McDowell with respect to the role of clothing in these gendered performances. However, in Colls's (2004) examination of emotions and women's experiences of clothing consumption, she points out how 'work which has considered women's relationships with clothing and fashion more generally have tended to dichotomize women's emotional connections with clothing as "positive" (playful, performative and celebratory) or "negative" (enslaving, patriarchal or indulgent)' (584). She moves on to argue how such readings emphasise 'the product of emotional sensation, i.e. positive and negative emotions, rather than how emotions are produced and experienced' (584). It would be possible to understand Emma's account in terms of the 'negative' emotions she experiences while walking in high heels. However, as Colls argues, this would tell us little about the emergent and produced nature of these emotions.

What Emma's account demonstrates is how emotion is used as a resource in her response to being asked about things that affect her 'feelings' of personal safety. The discursive context is already oriented to what it is to feel. Emma attends to the 'what' in terms of 'how' she feels about her night out, her clothes, her state of mind and body, herself, her job, her relations with strangers, variabilities of empowerment in relation to contextual significance of her femininity. All this is made available as a consequence of an assessment of her enjoyment of wearing shoes with heels in contrast to flat shoes. Emma formulates her feelings of safety on foot in terms of what it is to look feminine or not. She qualifies this as a matter of footwear facilitating her capacity to run. The right shoes combined with athletic prowess are what give her a sense of safety ('I definitely feel safer when I'm not looking so feminine, you know shoes I can run in, and I'm also a fast runner so that gives me a sense of safety'). She elaborates the significance of shoes in terms of what she normally wears and then contrasts that with her enjoyment in wearing 'heels'. She cuts short a further assessment of what she 'really hate(s)'. The formulation of the affective significance for her of shoe wearing occasions a reminder. This too is discursively accomplished – 'actually that's something'. The 'that's something' is then attended to not in terms of 'what' things make for feelings of personal safety when walking in general; rather she elaborates the significance of a particular journey home. She accomplishes this using a whole range of emotional and affective resources as she narrates the details of that journey.

As Sacks (1992) note, it is not sufficient to claim what he terms an 'entitlement to experience' as significant; it has to be shown to be so. The reporting of Emma's emotions is a primary resource in showing her entitlements to the frustrations of the night. Her story is set up as consequence of a 'really frustrating night out'. Thus, an evaluation in emotional terms sets the context for retelling the significance of this particular journey home. The shoes are not just heels; they are upgraded to 'high heel shoes'. Her embodied state is evaluated in terms of the extremes of fatigue as an affective state – 'I was really tired waiting for the bus in high heels'. Her shoes are also causally incorporated into that feeling of tiredness. They are situated further as the basis for thwarting her attempts to walk home – 'I started to walk home but I couldn't walk as I had these stupid shoes on'. Stupidity emotionally implicates her prior decisions to wear them and the shoes as objects in, and of, themselves. Emotion talk clearly matters in such accounts of an embodied pedestrian experience. However, it is possible to use Emma's account to probe a further range of concerns of relevance to the emotional geographies of walking.

Senses of place: anxieties and unease of being 'in' and 'out of' place

What Emma wears has an effect on the spatialities of her mobility as a pedestrian. Shoes combining with her athleticism do, as clearly indicated in her account, give rise to her sense of safety. Such safety is also a matter of the command she can exercise over the expansion of spatiality when the occasion demands. Her experience claims can be argued to reflect the shrinking and expanding of people's spatialities discussed in Chapter 4. Massey's (2001) example of her ageing father in the Wythenshaw area of Manchester draws attention to the ways in which spatialities of walking transform a person's sense of place:

> The spatiality of the very ordinary practice of walking to the shops is utterly transformed. And with it, your construction of this place. Your knowledge of it shifts. You don't look up to see the trees, or walk briskly through bracing air: you're having to concentrate on your feet. Your spatiality is closed down. Place is experienced, known, and thus made by embodied practices such as these. (464)

In like terms, Emma's 'sense of safety' in place is directly related in her account to the spatialities constituted in terms of which shoes she may or may not be wearing. Similarly, the implications of her sense of tiredness in 'waiting for the bus in high heels' relate to her emergent sense of place on her late night journey home. Her restricted mobility closes in her spatialities. As already described, it puts her in relation to people, circumstances, and 'stuff' that she would rather not be. Such 'stuff' is what configures the implications of her closed in, or contracted, spatiality in relation to her emerging sense of place concerning that journey. Places she would rather not be. A sense of place that imbricates her life in general at that time. A life potentially closed in. A restricted life in terms of economics and reported lifestyle all made relevant in the sense of place that she identifies as being 'so beyond this now'.

In their writings on 'emotions and social life', Bendelow and Williams (1998) point to the centrality of emotions to human experience. Davidson and Milligan (2004) also state how 'our emotions matter' as they go onto argue that 'they have tangible effects on our surroundings and can shape the very nature and experiences of our being-in-the-world. Emotions can clearly alter the way the world is for us, affecting our sense of time as well as space' (524). Emma's account draws attention to the emotional dimensions that can be associated with pedestrian movement and how they relate to her sense of space, yet Bondi et al. (2005) also point to the 'interconnected location of emotions in people and place' (5).

What makes for a sense of safety was clearly a concern in the extract from Emma's interview. Work on fear and public space has traditionally been approached via concerns with distinct gendered mobility patterns and the spatial tactics that individuals adopt as they encounter, avoid, and negotiate fear while walking, with the heterogeneity of these experiences acknowledging the 'multiple things and relations' that shape that fear (see for example Koskela and Pain, 2000; Pain, 2000). However, the analysis of the discursive accomplishment of emotion and affect in relation to place can provide a more nuanced approach to 'geographies of fear'. Of particular interest is how issues of anxiety, unease, and being 'out of place' (Cresswell, 2004) are occasioned in terms of emotional and affective claims. This in turn makes it possible to examine how senses of place are made a relevant concern and to provide evidential detail on the contradictory dimensions of urban life in terms of both the pleasure and anxieties experienced in the city (Wilson, 1991). For example, Lucy's diary entry makes visible the occasioning of feeling out of place with the rhythmic qualities of walking:

> I'm not walking that quickly, but I'm not hanging around either – I don't know Stratford very well, and I feel a little uneasy walking on my own here.
>
> (diary – Lucy, Canonbury resident, Islington)

Lucy makes relevant in this diary entry her unease of walking alone in a place she does not know well. She accomplishes this in terms of the way she herself might be held accountable in an unfamiliar place in terms of the pace and purposefulness of her walking. She is 'not walking quickly' but equally 'she is not hanging around'. 'Not hanging around' both qualifies her pace but it also entails a sense of purpose. It is that rhythmic combination of pace and purpose, that if not occasioned in relation to place would render her accountable, as someone who is 'out of place' (Cresswell, 2004). This combination is accounted for in terms of not knowing the area well. Lucy's account therefore makes available how her unease relates to her sense of place in terms of the ways it is 'provoked by being in a relatively unfamiliar place' (Urry, 2005: 77). Similarly, the next diary entry from a London Fields resident also makes relevant pace and purpose in his description of a particular late night walk home (see Chapter 5 on walking at night) and in so doing sets up the context for his anxiety:

> 02.30 – Left a friend's house in Shakespeare Walk after drinks and political conversation. Walked about .75 miles home, fast and purposeful – slightly anxious because it was so late/early.
>
> (diary – Dave, De Beauvoir Town resident, Hackney)

The time header for this diary entry is 2.30 in the morning. Dave attends to any accountability for being out late, as he puts it – so late it could be constructed as early. This has not been a night 'on the tiles',[4] he has been with friends, in their house. The implication is that time has slipped by, not just through drinking, rather in terms of what he makes a reportable feature of his entry – 'political conversation'. He has not a long journey home ('about .75 miles'). He describes it as 'fast and purposeful'. This is no simple reporting of speed as in the precise marking of distance ('.75 miles'). Pace is paired with a description of the rhythmic character of his walking at that time of night ('fast and purposeful'). It is the situated context of both the production of this diary entry and in the claims to the experience that are made; that rhythm in pace can be used to mark and account for what it is to report that he was 'slightly anxious because it was so late/early'. However, what Dave's entry also illustrates is how 'negotiating the city is a very different proposition in the hours of darkness, potentially involving heightened sensations of anxiety and excitement' (Hubbard, 2005: 120; see also Beaumont, 2015; Gallan, 2015; Sandhu, 2007; and Shaw, 2014, on the city at night). Hubbard (ibid) moves on to point out that for some:

> ... the apparently risky nature of the night-life is part of its appeal, and some may actively revel in encounters with difference or immersion in potentially risky or unpredictable situations. For most though the challenge is to negotiate these pleasures and dangers, using practical knowledge of the city to avoid situations that they would rather not deal with while seeking out forms of pleasure and stimulation. (120)

It is this everyday 'practical knowledge' (see Middleton, 2011), which continued to emerge in the context of exploring the emotional geographies of walking. For example, a diary entry describes Lucy's walking activity around Chinatown in central London:

> Couldn't find cab, bus would have taken too long. Strange smells as I walk through Chinatown and lots of traffic noise. Not exactly pleasant, but I do feel like I'm in the thick of it. I feel comfortable walking round this area – I know it pretty well.
>
> (diary – Lucy, Canonbury resident, Islington)

In contrast to Lucy's diary entry detailing her walking in Stratford she claims that the circumstances of her walk in China Town make her 'feel like I'm in the thick of it'. She feels 'in place'. Her claiming of awareness of both difference ('strange smells') and sensory excess ('lots of traffic noise') are precisely the sorts of features that make for feeling out of place. However, her comfort in this area is warranted in terms of her 'practical knowledge' of knowing this area 'pretty well'. It is this practical knowledge that frees her to actively exercise a 'right to mobility', discussed in the previous chapter, enabling her to engage with the sensualities of this particular place.

Bondi et al. (2005) argue that 'emotions are produced in relations between and among people and environments' (3) and 'that it may be productive to think of emotions as intrinsically relational' (7). The above discussion has highlighted the importance of how people use both 'sensual' and 'emotion talk' in their accounts of their embodied pedestrian experiences. These accounts have revealed how people discursively attend to their actions and how walking is a heterogeneous assemblage of embodied practices, sensual knowledges, affectual relations, and spatio-temporal configurations (Middleton, 2010). It is only by conceptualising walking as being socially and materially co-produced that the significance of the emotional geographies of everyday pedestrian practices becomes apparent. The data discussed has been drawn from in-depth interviews and accounts from walking diaries. In what follows I turn specific attention to these walking methods, and mobile methodologies more generally, and the associated challenges of doing research that engages with the embodied practices and emotional significance of everyday urban walking.

Walking methods

Over the last 15 years there has been a proliferation of writings, seminars, workshops, and conferences that consider mobile methodologies and more specifically walking as a research method. Much of the interest in walking methods relates to using walking as a way to explore, engage with, and understand space and place and includes a fascinating range of contributions from walking artists, activists, and academics. An overarching theme emerging from this work is concerns with embodiment and how the body can be used as a research tool. However, a further dominant presence throughout these engagements with walking as a method is the undifferentiated nature of walking itself, with the implicit assumption deeply embedded of a non-disabled, unencumbered, individual walking subject. Along with some others (see also Parent, 2016; Springgay and Truman, 2018, 2019), I have become increasingly troubled with the ways in which walking methods are frequently underpinned by a series of assumptions of both researcher and participant.

As Cresswell (2006) reflects upon how we conceive of mobility in terms of how 'it is absent the moment we reflect on it' and '... has passed us by' (58), he raises some significant methodological issues in terms of what it is to research movement and the types of methods appropriate for doing so. What methods can be drawn upon to capture something that has always 'passed us by' and is supposedly 'absent' from the moment it is reflected on? In Sheller and Urry's (2006) frequently cited proposition of a 'new mobilities paradigm' they drew specific attention to a range of methods they consider suitable for mobilities research. These methods include observations of movement, mobile ethnography, time-space diaries, cyber-research, and the use of material objects such as 'photographs, letters, images, souvenirs' (218) as resources for performances of travel memories.

To date, mobile methods have been adopted in research engaged with wide-ranging concerns including multiculturalism (Wilson, 2011), urban regeneration (Ricketts-Hein et al., 2008), the lived experiences of higher education students (Holton and Riley, 2014), commuting practices (Jirón, 2010), disability and occupational health (Butler and Derrett, 2014), and visual impairment (Macpherson, 2009). However, perhaps the most frequently drawn upon mobile method is that of the 'go-along'. The term 'go-along' was first coined in the work of Kusenbach (2003) on urban neighbourhoods. For Kusenbach, adopting an approach of moving along with research participants was 'particularly suited to explore: environmental perception, spatial practices, biographies, social architecture and social realms' (455). Go-alongs have been subsequently adopted as a means of engaging with many different mobile practices ranging from cycling (Spinney, 2009, 2011) to driving (Laurier, 2010). However, one of its most common manifestations is in the form of pedestrian go-alongs and their emergence as a dominant method in researching people's experiences with space and place (see for example Anderson, 2004; Bates and Rhys-Taylor, 2017; Kelly et al., 2011; O'Neill and Roberts, 2020; Ricketts-Hein et al., 2008; Silver et al., 2020).

In Spinney's (2015) critical examination of mobile methods, he considers the potential contribution of bio-sensing technologies being used in conjunction with go-along approaches. In providing the broader context for the paper, Spinney acknowledges the diversity with which mobile methods have been interpreted across a range of different work alongside the different rationales for their adoption. However, in doing so, he raises particular issues concerning representation. First, he proposes that those adopting go-along methods do so as a means of understanding 'the fleeting, relational and felt aspects of mobility which resist representation' (232). The second, related, point is that go-alongs 'are geared towards enabling participants and researchers to participate and reflect on practices rather than any subsequent representation of practice' (233). Spinney is not alone in privileging mobile methods such as go-alongs as more appropriate than what are frequently considered more traditional social science methods, such as static in-depth interviews, to capture the embodied experiences of movement. His position relating to issues of representation reflects approaches that have stemmed from Thrift's (2004, 2007) concern with 'non-representational theory'. Spinney (2015) argues that this theoretical orientation stems from concerns with the significance of everyday practices and the 'sensory and affective experience – that "overflow" our ability to apprehend and represent them through language' (232).

For Merriman (2014), using mobile methods to engage with the mobilities of research participants 'in more direct and multi-sensuous ways' is based on 'a rather problematic assumption that these methods enable the researcher to more accurately know and represent the experiences of their research subjects' (8). There are several examples of recent research adopting walking methods that exemplify these theoretical tendencies and assumptions in mobilities research. For example, in Wolifson's (2016) exploration of nightlife spaces in the Sydney suburb of Surry Hills, go-alongs combined with 'talking whilst walking' (Anderson, 2004) were adopted as a means of understanding the experiences and encounters in these spaces in relation to the broader context of night-time economic planning. For Wolifson, this approach enabled her to 'remain cognisant of the different sensory levels of complex conscious and unconscious relationships between self and landscape' (8). She also emphasises Hill's (2013) understanding of walking as a method which facilities 'powerful recollections' and a tactile engagement with the world (391).

Evans and Jones (2011) utilised walking interviews as a means of exploring notions of place in the context of local residents' experiences of a redevelopment scheme in the Digbeth district of Birmingham. The research aimed to understand the relationship between what people were saying and where they were saying it and to explore the differences between sedentary and mobile interviews. They propose that 'focusing on words and location means that some of the more embodied characteristics of interview practice are lost' (851). As Latham (2003) explores how the 'performative ethos can inform and invigorate the human geographic imagination' (1993), he expresses similar perspectives relating to how embodied social practices are researched. It is proposed as particularly troubling that 'talk is made to stand in for all the complexities and subtleties of embodied practice' and how 'the cultural geographer can – it would appear – speak the world to truth through asking her or his research subjects the right questions, and then quoting them back with fidelity in their research reports' (1999).

While this growing/renewed interest in walking methods across the social sciences is relatively recent, drawing upon pedestrian practices as a method or mode of engagement is far from a new approach. For example, Solnit (2014) explores the relationships between the body and walking through her discussion of peripatetic philosophers. She considers the carefree wanderings of Rousseau and others as a means of emphasising the centrality of embodied understandings of pedestrian practices. For, as she claims, 'walking shares with the making and working that crucial element of engagement of the body and the mind with the world, of knowing the world through the body and the body through the world' (29). However, Solnit's discussion also brings sharply into focus the romanticised and privileged version of walking that circulates through much literary and philosophical writings. Artistic engagements with walking and the adoption of pedestrian movement as a thinking/intellectual practice can be read through a lens of privilege. For many people, no such space exists for them to adopt such methods and forms of artistic expression. Although, as I have argued earlier in the book, these romanticised versions of walking, that are only experienced from a position of privilege, are prevalent across other arenas including those relating to transport and health. Springgay and Truman (2018) draw attention to the value of walking as a research methodology, yet also urge caution in highlighting how 'walking is never neutral' and that 'it is crucial that we cease celebrating the White male flâneur, who strolls leisurely through the city, as the quintessence of what it means to walk' (14).

Keinänen and Beck (2017) critically examine the gendered experiences of walking for intellectual work and propose a research agenda on gender, walking, and thinking. Their work raises important questions around gender and the roles of the body/embodiment in intellectual work. However, an acknowledgement of 'other' voices is largely absent from the paper. How does the embodied practice of walking and thinking intersect with social identities associated with race, class, or disability? Furthermore, as I have highlighted, much of the debate around walking methods, and mobile methodologies more generally, concerns whether the sensory, emotional, and affective dimensions of mobile experiences are unspeakable and whether traditional social science methods such as interviews and focus groups are inadequate at capturing mobile experiences and practices (Fincham et al., 2010). Yet, what remains underexplored is, in the words of Butler, that 'nobody goes for a walk without something to support that walk' (Taylor et al., 2010) This point relates to concerns with interdependency and the accessibility of walking methods to different groups. A notable exception is Macpherson's (2016) writing on using walking methods in landscape research. Macpherson challenges the 'methodological orthodoxy' of walking methods to 'open up new spaces of disclosure, aid rapport and enable new knowledge of landscape ...' (426). In highlighting the limits of walking methods for understanding our experiences of landscape, she argues that 'walking and walker's bodies bring with them their own politics, cultures, histories, habitual responses and lived experiences that must be taken into account (Edensor, 2000, 2010; Heddon, 2012; Myers, 2011)' (426). More recently, Springgay and Truman (2019) have pushed this point further in contending that critical walking methodologies must move beyond concerns with health or innovative methods and pay close attention to walking in relation to anti-ableism, anti-racism, and anti-colonialism (see also O'Neill and Einashe, 2019).

In what follows I will draw upon research on the everyday mobilities of VI young people as a means of critically engaging with walking methods in relation to concerns with accessibility and interdependency. I argue that far more attention is needed on these important, yet frequently ignored, dimensions of adopting walking methods. In particular, I focus on this group's experience of adopting a mobile methodology, which included walking, in research concerned with their independent mobility and visual impairment. In doing so, I bring what Warren (2017) refers to as the 'demands of the method' (802) to the centre of analysis while critically reflecting upon the politics emerging from such a move.

Everyday VI mobilities, walking methods, and interdependent research relations

This work emerged from a dialogue with the Royal London Society for Blind Children (RSBC)[5] relating to the type of research informing their thinking on independent mobility. More specifically, there were concerns that this research was predominantly quantitative with assumptions being made by sighted adults as to what independent mobility actually was. As such, a project was developed in collaboration with the charity focusing on transport infrastructures and the everyday mobility practices/experiences of VI young people. When the dialogue with the RSBC began, they had just launched the Youth Manifesto. This document was produced by the RSBC Youth forum that was set up to 'act as a megaphone for vision impaired young people in London and the South East and to unearth potential solutions to challenges we face, such as employment, transport and accessible technology' (RLSB, 2014). Transport and accessibility form a key part of this manifesto with particular attention being drawn to some of the challenges and potential solutions surrounding both Overground and Underground trains and bus transit. The aim with the project was to complement the work of the Youth Forum with research that explored some of these transport issues in greater depth. In particular, the study aimed to explore the relationship between travel infrastructures and VI young people's everyday lives in the context of generating recommendations about how transport could be better designed and managed for these groups in the future.

Methodologically, a key aspect of the research was developing what could loosely be termed a participatory video method. Parent (2016) urges us to not only think about uneven mobilities but also about uneven methods and associated power relations. She calls for researchers to not simply go beyond disability but to 'go-along' with disability and disabled people (see also Moles, 2019; Palmgren, 2018, on the power dynamics of mobile methods including walking methods). The decision to draw upon such an approach stemmed from a desire to not only attempt to negotiate some of the complex power relations embedded in mobilities research with VI young people who are 'moving into' adulthood (see Nijs and Daems, 2012) but also from a wish to develop some sort of further critical engagement with mobile methods (see Merriman, 2014). Participatory video has been effectively used in research on mental health (Parr, 2007) and those living with long-term health conditions (Bates, 2014), yet video as a research tool has been used less frequently in research on VI.

There are clearly tensions in the use of video as a methodological tool for research on VI as it can be positioned as over-privileging the visual sphere. However, while participants had different levels of vision, the audio-visual recordings were also of significant value for engaging with a range of VI experiences. For example, the audio-visual data proved to be a key resource in highlighting the significance of senses other than sight and their emergence as 'orientation cues' (Saerberg, 2010) in everyday mobility practices. This brought to the fore the significance of urban textures and materalities; the materiality of the camera, buttons, and harness; changes in light; proximity to other bodies; and emotions and affects such as anxiety, fear, accomplishment, and pride. These findings challenge the assumption that VI people are not able to relate to the visual sphere. As the research had been funded as a pilot project, we worked with a relatively small number of participants totalling seven[6] with notions of independence/interdependence emerging as important methodological considerations.

It became evident as the research progressed that the notion of independence was contested by the methodological approach to the research. For example, travelling with a camera, capturing footage and images to show and share with others (researchers/policy-makers/friends and family) as 'evidence' is intrinsically interdependent in terms of an entanglement of different technologies and people. The initial research design was framed around a 'participant led' approach. However, the participants required more guidance than expected in terms of the video content, fieldwork, and literally on the 'go-alongs' that drew us into more interdependent relations with regard to the fieldwork process. We therefore began to realise that we, as researchers, had in fact made a series of assumptions around notions of 'independent mobility', which were embedded in the methodology. Despite participants being able to lead and direct the research towards their own interests and experiences, these emergent fieldwork issues further highlight the normative and assumed dimensions associated with what it is to be 'independently mobile'.

Schwanen et al. (2012) reflect upon independence and mobility in later life through examining the everyday mobility practices of community dwelling older people in the UK. They argue for a broader conceptualisation of independence that recognises the embedded assumptions of what it is to be independent and the complex ways in which independence is a relational accomplishment. In developing this argument, they suggest that 'both independence and mobility are fabricated out of myriad relations with and dependencies on bodies, technologies, infrastructures, social networks and other forms of materiality (as well as social conditions)' (1320). Schwanen et al. do not explicitly use the term 'interdependence' but their analysis is suggestive of such a concept. As it became clear that interdependence had become woven into our research methodology, this in turn caused us to question some of the assumptions underlying our initial intention for the research to be 'participant led'.

In response to the histories of medicalised and occularcentric approaches to VI within geography, recent research has experimented with the use of multi-sensory or mobile walking methods (Büscher et al., 2010) to gain more embodied insights into VI mobility experiences (Hammer, 2013; Macpherson, 2009; Porcelli et al., 2014; Worth, 2013). For example, in her research with VI adult walkers in the Lake District, Macpherson (2009) applied the go-along approach (Kusenbach, 2003), to her experience of acting as a sighted guide for her participants. In describing how her body took on a role of mediating touch between the landscape and blind walkers, the research demonstrates how the theme of interdependency plays out in researcher-participant relations. Macpherson's experiences capture an interesting dynamic between researcher and participant, in which a certain pathway is followed and navigated as a result of a relational exchange between the two. Such a dynamic troubles the (often binary) separation between 'participant' and 'researcher', destabilising the notion that research can be solely participant or researcher 'led', offering a more interdependent and 'messy' (Law, 2004) understanding of participation and agency within research.

Parent (2016) proposes the go-along method as a way of redressing unequal power relations, which devalue disabled experiences as ways of knowing and moving through the world. These concerns are also present in Hammer's (2013) research with VI women in Israel. Hammer argues that through engaging in 'physical intimacy' by walking with and guiding participants in public space, she is able to share something of their experience, becoming aware of non-visual sources of knowledge such as texture, liquidity, temperature, shape, and sound, 'recognising haptic, olfactory, and gastronomic experiences'. She observed how participants interacted with her and her research tools, often critiquing and questioning her methods and positionality as a researcher, causing her to shift and question her own focus: 'I had to conduct this research not only in order to better understand the life experiences of a blind woman, but to learn what it means to be a sighted woman, and more specifically, what it means to experience sightedness'.

The work of Macpherson, Parent, and Hammer demonstrates the ways in which researchers become drawn into interdependent relationships with participants in the field through the use of walking methods, raising important questions around ethics, care, and exchange (Gunaratnam, 2012). Their work points to the possibility for different ways of knowing through the use of reflexive, multi-sensory, and embodied mobile methods (Pink, 2009). This reflexivity is central to the work of Lomax (2015) and her use of creative visual research with children, in which she calls on researchers to be attentive to their own research process, recognising that the voices and experiences of participants are negotiated in and through that process. As I outlined in the preface and introductory chapter, my own experiences of urban walking have inevitably shaped the lenses through which I have engaged with the material presented throughout this book. I now turn attention to the experiences of the participants of the VI research while acknowledging the importance of our own research process (see Lomax, 2015). I also draw attention to the significance of other research relations to participants' experiences.

Each participant of the project being discussed here engaged with and harnessed the research methods in uniquely different ways, incorporating them into their own 'styles' and ways of moving through and being in the city. This allowed us to witness and participate in the mobile methods of participants. Through this we realised that, rather than being a static state or destination to arrive at, independence was something constantly being mediated through these everyday methodologies. One example from the video footage showed how Katerina struck up conversation with a member of the public sitting next to her on the train, asking if the person could help her turn on the GoPro and check it was recording. In the proceeding videos, this member of the public appears in different stations and locations across London, evidently assisting Katerina on her journey, and demonstrating how Katerina used the GoPro to mobilise the help and support she needs to get around in unfamiliar places.

RESEARCHER: So what would you do if you don't feel like you know an area?
KATERINA: I just ask the public to help me. I rely on the public a lot.

Another method for gaining and sharing knowledge of unfamiliar routes through the city emerged in conversation with Sophie, Salih, and James – all of whom described methods for supporting peers with mobility challenges. This peer support could come in the form of making videos and audio recordings of journeys to share with other VI young people, helping friends to navigate different technologies (such as the GoPros), making YouTube videos and channels, or literally guiding friends and partners in peer groups. James told us about how he had on occasion guided up to six VI friends (plus guide dogs) in formation on a journey, describing in detail his tactics and methods for doing so, as well as the importance of sharing responsibility among the group. Despite the wide variation in participants' mobility methods, they all shared a common theme/strategy of interdependence, perhaps revealing how interdependence is itself a mobile method, harnessed in different ways to mediate and achieve independence in different moments. However, the accessibility of the method warrants closer attention.

Walking methods and accessibility

I have already noted the critique of mobile method approaches being based on the problematic assumption that they somehow give more 'authentic' access to experience (see Merriman, 2014). Yet, such a positioning of mobile methods negates a further, more important, dimension of access in terms of the accessibility of the method for research participants. For example, in interviews and video footage collected by Sophie, the 'fear' and 'panic' at getting lost, or left on one's own, came up again and again, creating stress and posing a further barrier to mobility and access. In the extract below this was exacerbated by the demands of the camera element of the method:

SOPHIE: Because obviously we got a little bit lost in that one, trying to find our way around. That for me is a fear, I hate getting lost, I freeze up, I panic, I get upset, I can't – if I'm with friends and they've got like a logical head on their shoulders I can do it, two heads are better than one. But if I'm on my own and there's no one around to help me, I can't stay logical. I just freeze, I can't deal with it.

RESEARCHER: Thinking about the methodology as well and the camera, did that feeling of panicking and freezing with just trying to use this GoPro camera, did that feel – was there any parallel between that and getting lost?

SOPHIE: It kind of adds a lot more stress to it, because you've got to think, have I turned it on correctly, has it made the right amount of beeps, have I got it on myself? For me personally it adds another level of – not in another way, but another level of stress to the journey in effect.

For Sophie, the experience of being alone is almost synonymous with being lost, or feeling anxious, and stressed. The powerful affects of fear and terror that she describes here are also linked to a bodily sense of 'freezing up'. Several participants shared similar embodied reactions to concerns with wayfinding and getting lost. Their experiences sit in stark contrast to the long association between walking methods and an intentionality to lose one's way as a means of engaging with the city. This is most prevalent in the work of psycho-geographers emerging from the Situationist movement (see Chapter 3). Yet adopting these walking methods and being able to take up an invitation to 'get lost' in the city is a privileged one and a privilege relating to class, gender, and race that is infrequently recognised across much of the tradition of psycho-geography (although notable exceptions include the work of Morag Rose and the Loiterers Resistance Movement detailed in Morris and Rose, 2019).

As I discussed in Chapter 2, the assumptions around walking bodies are also reflected in broader concerns with wayfinding in cities, where the embodied experiences of differentiated mobile bodies are frequently neglected. Wayfinding has long been engaged with in disability and medical literature on visual impairment (see for example Golledge et al., 1996; Koutsoklenis and Papadopoulos, 2011; Manduchi and Kurniawan, 2011), yet a dominant discourse that feeds through much of medical literature relates to improving 'performance', through avoiding potential collisions, and increasing the 'efficiency' of walking speeds for VI people (see for example Clark-Carter et al., 1986). In other words, the overarching aim of such research is to seek ways in which walking speeds can be reached that match those of a non-disabled/sighted pedestrian. This is also reflected in the development of activities by organisations such as Wayfindr, a non-profit organisation aiming to create a benchmark in standards for accessible digital navigation. However, while there is certainly much value in these tools, there is an underpinning neoliberal demand that VI people are assimilated into the capitalist flows, rhythms, and ideologies of urban space. While there is now a well-established critique of such medicalised/neoliberal approaches to disability, far less attention has been paid to these concerns in relation to walking methods.

Sophie's struggles and frustrations coming into contact with the GoPro in the context of using a visual walking method demonstrate the need for care in the use of this technology as a methodological tool for research with different groups of people. It is possible to imagine other technological tools that have design features supposed to make walking easier and more pleasurable, such as FitBits and the Apple Watch,[7] prompting similar frustrations. The GoPro method effectively slowed Sophie down in her daily mobilities, revealing that the design of the GoPro is not accessible for some VI people. Despite our best attempts to support her with the camera and reassure her that she did not have to use it at all, the presence of the GoPro still brought her stress. This was an unanticipated limitation of the research design, which is deeply regretful. It is worth noting that several other participants used the GoPro as a method for sparking up conversations with others. As such, the GoPro mediated these encounters leading to what Moser and Law (1999) refer to as 'good passages'.

Castrodale (2018) notes how non-disabled researchers should be cautious about the assumptions they make in research design. He highlights how 'accessibility concerns relating to physical and attitudinal barriers represent a significant rationale shaping participants' desires to engage in particular research interview methods' (52). In line with Castrodale's concerns, I argue here that in adopting walking methods, greater care should be taken in considering the experiences of research participants, particularly around issues of method accessibility. As such, it is important to reflect upon and share the technical difficulties experienced as part of the project in order to contribute to further understanding of using these methods with young VI people as well as other groups. As I have suggested, the technology used as part of this research was not accessible to all participants, and some experienced various technical difficulties, for instance with the battery life, switching between settings, the physicality of the harness, buttons and clips, and connecting the device to a laptop to charge or upload pictures. A lot of technologies, and particularly cameras, are not designed with visual impairment or other disabilities in mind (Pullin, 2009). As Parent (2016) points out, this is particularly the case with the GoPro, originally developed for extreme action videography, and whose brand image assumes a non-disabled user/customer.

As a disabled researcher, Parent (2016) found creative ways to adapt the GoPro technology, which involved assistance either from her participants or strangers. Participants in our study similarly found ways to adapt the technology, with support – producing interesting and important results. This adaptation work challenges the assumption that VI people have no use or need for cameras or image-making technology. In particular, it shows how camera technologies are already part of their mobility methods (the allure of the GoPro was a key factor for some participants getting involved), and that video cameras can be understood more as a polysensory recording devices (Pink, 2009) and not simply visual technologies. At the same time, this adaptation work caused stress and difficulty for some participants. For others, the presence of the technology potentially made it harder for them to engage in the research project, thus highlighting why recruitment was so challenging, with the project taking much longer than anticipated.

In some cases, as the inaccessibility of the methods emerged, the conditions were created for an interdependent research practice. It simply was not possible for certain participants to navigate the adopted method without help. It was this interdependency that continued to feed into and inform the final stage of the research process. It had always been the intention to hold an end-of-project workshop which involved participants. However, the details of this were not finalised at the time we began data collection. As the fieldwork stage unfolded it became clear how important it was to maintain these research inter-relations as the edit and analysis began of the film footage. Castrodale (2018) emphasises 'ethics of interconnectedness' (45) in the use of mobile methods. In doing so, he argues 'that researchers have an ethical responsibility to research *alongside* disabled persons, to research *with* disabled persons, and not to engage in predatory research practices that advantage the research without reciprocation (Kitchin, 1999; Stone and Priestley, 1996; Zarb, 1992)' (45–46).

While the inaccessibility of methods was never intentional, and had several regrettable unintended consequences noted above, it also forced consideration of how we might continue to work alongside our participants in the analysis, editing, and production of the video data they had collected. The film workshops run at the end of the project were an attempt to attend to the power relations and politics of research design through a process of co-production (see Castrodale, 2018; see Middleton and Byles, 2019, for more detailed discussion). Yet, for me, the question remains as to how walking methods, and mobile methodologies more broadly, translate into wider policy and practice contexts which remain dominated by research that either sidelines or ignores the significance of the embodied and affective dimensions of pedestrian practices or work drawing upon a traditional toolkit of quantitative methods. This has implications for how walking in the city is understood insofar as many walking methods assume a non-differentiated walking body. Such research risks reproducing narrow and essentialised versions of urban walking, which fail to take into account that not all walking is the same.

Conclusion

I began this chapter with Judith Butler's question of 'what can a body do?' In addressing this question in the context of walking I have sought through this chapter to emphasise the centrality of the body to everyday pedestrian experiences. Considering the embodied and emotional geographies of walking enables us to critically examine many of the assumptions underpinning how walking is understood, positioned, and conceptualised. These range from the significance of the body being overlooked completely, to a lack of attention to bodily difference or the emotional and affective dimensions of walking bodies. Embodied experience is centre stage to many performative and artistic engagements with walking but much of this work emerges from a position of privilege in terms of having the time or resources to be able to undertake such work in the first instance. This risks producing very romanticised perspectives on walking, which do little to differentiate between different walking bodies. In seeking to challenge the assumptions embedded in these particular walking approaches, it is helpful to move to Butler's second concern of 'what supports a walk' in the context of the central conceptual thread of this book concerned with how walking is socially and materially co-produced. It is this focus that allows us to interrogate the inequalities emerging from how walking is adopted as a research tool and the significance of the frequently neglected emotional and affective dimensions of pedestrian embodied experience. This is best exemplified when we consider more closely how walking methods work in practice.

As discussed earlier, there is a tendency for walking methods to be adopted as a means of engaging with embodied experience. However, little explicit attention is given in this work to accounts of the experiences of those engaged with such methods. This is problematic as the accessibility, or 'demands' (Warren, 2017: 802), of the method remains invisible. This point extends to how walking methods are conceptualised and the need to broaden conceptualisations to allow space for differentiated walking bodies. I have argued here that it is through focusing on the embodied and emotional geographies of research participants' experiences which make visible the extent to which walking methods are accessible to different groups of people. Differentiated mobilities are an important, yet neglected, consideration in the adoption of walking methods and mobile methodologies more broadly. Furthermore, this is not simply a methodological concern but one which relates to mobility justice. This is an important issue and one to which I will return in the next, concluding, chapter. How walking is used as a research tool matters and the embodied and affective experiences emerging from how these approaches are adopted, interpreted, and implemented needs to be taken account of. As such, it is important to challenge existing engagements with walking which come from a particular position of privilege and fail to take account of differentiated mobile experiences on foot. When you start to consider the walking experiences of different individuals, groups, and social identities beyond a white, middle class, non-disabled, male, it is possible to interrogate these assumptions and begin to engage more closely with the emergent politics of these experiences.

Notes

1. The 'Examined Lives' documentary was directed by Astra Taylor and consists of interviews with eight philosophers speaking about the applicability of their work to contemporary life. The philosophers featured in the film are Kwame Anthony Appiah, Judith Butler, Michael Hardt, Martha Nussbaum, Avital Ronell, Peter Singer, Cornel West, Slavoj Žižek.
2. CASE studentships were implemented by the UK Government 'to make publicly funded academic researchers more responsive and relevant to the demands of research "customers" in business and government' with 'initiatives designed to promote so-called knowledge transfer between universities and non-academic organizations' (Demeritt and Lees, 2005: 128, 136).
3. This year-long study explored the everyday experiences of new parents in the city of Oxford within the broader context of the withdrawal of many key support services due to centrally imposed austerity measures. A photo diary-interview method was adopted to examine the experiences of 16 first-time parents.
4. An expression referring to a late night out at bars or clubs. Tiles is often taken to mean the tiles of the dance floor.
5. In 2017, the Royal Society for Blind Children and Royal London Society for Blind People (RLSB) joined together under the name of the Royal Society for Blind Children.
6. The final pack we gave to each participant included a GoPro camera, chest harness, cable to download footage, instruction manual, and laminated information card, which all participants carried with them in the event of being challenged about what they were doing. The card had details of the project, the researcher's contact details, and a key point of contact at Transport for London (TfL). When each participant was met to hand over the pack, we also conducted an accompanied 'go along' as means of familiarising them with the equipment and providing them with an opportunity to ask any questions. Each participant had the kit for a couple of weeks before we met for a follow-up interview to go through some of the footage they recorded of their everyday journeys.
7. Fitbit is a wireless, wearable technology that tracks physical activity, such as number of steps walked and measures data including heart rate, quality of sleep, and other bodily metrics associated with maintaining fitness levels. Apple Watch is a smart watch that, along with email and internet functions, is also able to track and measure fitness data.

References

Adams, M. & Guy, S. 2007. Editorial: Senses and the City. *The Senses and Society: The Senses and the City*, 2, 133–136.

Adams, M., Moore, G., Cox, T., Croxford, B., Refaee, M. & Sharples, S. 2007. The 24-hour city: Residents' sensorial experiences. *The Senses and Society: The Senses and the City*, 2, 201–215.

Adey, P., Bissell, D., McCormack, D. & Merriman, P. 2012. Profiling the passenger: Mobilities, identities, embodiments. *Cultural Geographies*, 19, 169–193.

Alayo, J. 2002. Measuring pedestrian accessibility in urban environments, Proceedings of the 2002 European Transport Conference, PTRC, London.

Anderson, B. & Harrison, P. 2006. Commentary: Questioning affect and emotion. *Area*, 38, 333–335.

_____. 2010. *Taking-place: Non-representational theories and geography*, Farnham, Ashgate.

Anderson, J. 2004. Talking whilst walking: A geographical archaeology of knowledge. *Area*, 36, 254–261.

Ash, J. & Simpson, P. 2016. Geography and post-phenomenology. *Progress in Human Geography*, 40, 48–66.

Barker, J. 2011. Manic mums' and 'distant dads'? Gendered geographies of care and the journey to school. *Health & Place*, 17, 413–421.

Bates, C. 2014. Intimate encounters: Making video diaries about embodied everyday life. In: Bates, C. (ed.) *Video methods: Social science research in action*, London, Routledge, 20–36.

Bates, C. & Rhys-Taylor, A. 2017. *Walking through social research*, New York, Routledge.

Beaumont, M. 2015. *Nightwalking: A nocturnal history of London, Chaucer to Dickens*, London, Verso.

Bendelow, G. & Williams, S. J. 1998. *Emotions in social life: Critical themes and contemporary issues*, London, Routledge.

Bondi, L. & Davidson, J. 2011. Lost in translation: A response to Steve Pile. *Transactions of the Institute of British Geographers*, 36, 595–8.

_____. 2016. *Emotional geographies*, London/Abingdon, Routledge.

Bondi, L., Davidson, J. & Smith, M. 2005. Introduction: Geography's 'emotional turn'. In: Davidson, J., Bondi, L. & Smith, (eds.) *Emotional geographies*, Farnham, Ashgate, 1–18.

Borer, H. & Borer, H. 2013. *Structuring sense: Volume III: Taking form*, Oxford, Oxford University Press.

Boyer, K. & Spinney, J. 2016. Motherhood, mobility and materiality: Material entanglements, journey-making and the process of 'becoming mother'. *Environment and Planning D: Society and Space*, 34, 1113–1131.

Büscher, M., Urry, J. & Witchger, K. 2010. *Mobile methods*, London, Routledge.

Butler, M. & Derrett, S. 2014. The walking interview: An ethnographic approach to understanding disability. *Internet Journal of Allied Health Sciences and Practice*, 12, 6.

Castrodale, M. A. 2018. Mobilizing dis/ability research: A critical discussion of qualitative go-along interviews in practice. *Qualitative Inquiry*, 24, 45–55.

Clark-Carter, D. D., Heyes, A. D. & Howarth, C. I. 1986. The efficiency and walking speed of visually impaired people. *Ergonomics*, 29, 779–789.

Clement, S. & Waitt, G. 2017. Walking, mothering and care: A sensory ethnography of journeying on-foot with children in Wollongong, Australia. *Gender, Place & Culture*, *24*, 1185–1203.

Colls, R. 2004. Looking alright, feeling alright': Emotions, sizing and the geographies of women's experiences of clothing consumption. *Social & Cultural Geography*, *5*, 583–596.

Crary, J. 1999. *Suspensions of perception: Attention, spectacle, and modern culture*, Cambridge, MA, MIT Press.

Cresswell, T. 1999. Embodiment, power and the politics of mobility: The case of female tramps and hobos. *Transactions of the Institute of British Geographers*, *24*, 175–192.

_____. 2004. Defining place. In: *Place: A short introduction*, Malden, MA, Blackwell Ltd, 12.

_____. 2006. *On the move: Mobility in the modern Western world*, New York, Routledge.

Curti, G. H., Aitken, S. C., Bosco, F. J. & Goerisch, D. D. 2011. For not limiting emotional and affectual geographies: A collective critique of Steve Pile's 'Emotions and affect in recent human geography'. *Transactions of the Institute of British Geographers*, *36*, 590–594.

Davidson, J. & Milligan, C. 2004. Embodying emotion sensing space: Introducing emotional geographies. *Social & Cultural Geography*, *5*, 523–532.

Dawney, L. 2011. The motor of being: A response to Steve Pile's 'Emotions and affect in recent human geography'. *Transactions of the Institute of British Geographers*, *36*, 599–602.

Degen, M. 2014. The everyday city of the senses. In: Paddison, R. & McCann, E. (eds.) *Cities and social change*, Los Angeles, Sage, 92–111.

Degen, M., Desilvey, C. & Rose, G. 2008. Experiencing visualities in designed urban environments: Learning from Milton Keynes. *Environment and Planning A*, *40*, 1901–1920.

Demeritt, D. & Lees, L. 2005. Research relevance, 'knowledge transfer' and the geographies of CASE studentship collaboration. *Area*, *37*, 127–137.

Doughty, K. 2013. Walking together: The embodied and mobile production of a therapeutic landscape. *Health & Place*, *24*, 140–146.

Dowling, R. 2000. Cultures of mothering and car use in suburban Sydney: A preliminary investigation. *Geoforum*, *31*, 345–353.

Edensor, T. 2000. Walking in the British countryside: Reflexivity, embodied practices and ways to escape. *Body & Society*, *6*, 81–106.

_____. 2010. Introduction: Thinking about rhythm and space. *Geographies of Rhythm: Nature, Place, Mobilities and Bodies*, 1–18.

Edwards, D. 1997. *Discourse and cognition*, London, Sage.

Edwards, E. & Bhaumik, K. 2008. *Visual sense: A cultural reader*, Oxford, Berg.

Elden, S. 2004. (translated by S. Elden and G. Moore). Rhythmanalysis: An introduction. In: Lefebvre, H. (ed.) *Rhythmanalysis: Space, time, and everyday life*, London, Continuum, vii–xv.

Evans, J. & Jones, P. 2011. The walking interview: Methodology, mobility and place. *Applied Geography*, *31*, 849–858.

Feld, S. 2005. Places sensed, senses placed: Towards a sensuous epistemology of environments. In: Howes, D. (ed.) *Empire of the senses: The sensual culture reader*, Oxford, Berg.

Fincham, B., McGuinness, M. & Murray, L. 2010. *Mobile methodologies*, Basingstoke, Palgrave Macmillan.

Gallan, B. 2015. Night lives: Heterotopia, youth transitions and cultural infrastructure in the urban night. *Urban Studies*, *52*, 555–570.

Golledge, R. G., Klatzky, R. L. & Loomis, J. M. 1996. Cognitive mapping and way-finding by adults without vision. In: Portugali, J. (ed.) *The construction of cognitive maps*, Dordrecht, Springer, 215–246.

Goodwin, M. H. 2006. *The hidden life of girls: Games of stance, status, and exclusion*, Malden, MA, Oxford, Blackwell.

Goodwin, S. & Huppatz, K. 2010. *The good mother: Contemporary motherhoods in Australia*, Sydney, Sydney University Press.

Grosz, E. 1994. *Volatile bodies: Toward a corporeal feminism*, Bloomington, Indiana University Press.

Gunaratnam, Y. 2012. Learning to be affected: Social suffering and total pain at life's borders. *The Sociological Review*, *60*, 108–123.

Hammer, G. 2013. "This is the anthropologist, and she is sighted": Ethnographic research with blind women. *Disability Studies Quarterly*, 33.

Heddon, D. 2012. Turning 40: 40 turns: Walking and friendship. *Performance Research*, *17*, 67–75.

Heddon, D. & Turner, C. 2012. Walking women: Shifting the tales and scales of mobility. *Contemporary Theatre Review: Site-specificity and Mobility*, *22*, 224–236.

Hill, L. 2013. Archaeologies and geographies of the post-industrial past: Landscape. Memory and the spectral. *Cultural Geographies*, *20*, 379–396.

Holton, M. & Riley, M. 2014. Talking on the move: Place-based interviewing with undergraduate students. *Area*, *46*, 59–65.

Horton, D. 2005. Book review: Henri Lefebvre, Rhythmanalysis: Space, time and everyday life. *Time and Society*, *14*, 157–159.

Hubbard, P. 2005. The geographies of 'going out': Emotion and embodiment in the evening economy. *Emotional Geographies*, 117–134.

_____. 2006. *City*, Abingdon/New York, Routledge.

Ingold, T. 2000. *The perception of the environment: Essays on livelihood, dwelling and skill*, London, Routledge.

Ingold, T. 2004. Culture on the ground: The world perceived through the feet. *Journal of Material Culture*, *9*, 315–340.

Jaffe, R., Dürr, E., Jones, G. A., Angelini, A., Osbourne, A. & Vodopivec, B. 2020. What does poverty feel like? Urban inequality and the politics of sensation. *Urban Studies*, *57*, 1015–1031.

Jensen, M. T. 2018. Urban pram strolling: A mobilities design perspective. *Mobilities*, *13*, 584–600.

Jirón, P. 2010. Mobile borders in urban daily mobility practices in Santiago de Chile. *International Political Sociology*, *4*, 66–79.

Keinänen, M. & Beck, E. E. 2017. Wandering intellectuals: Establishing a research agenda on gender, walking, and thinking. *Gender, Place & Culture*, *24*, 515–533.

Kelly, C. E., Tight, M. R., Hodgson, F. C. & Page, M. W. 2011. A comparison of three methods for assessing the walkability of the pedestrian environment. *Journal of Transport Geography*, *19*, 1500–1508.

Kitchin, R. 1999. Morals and ethics in geographical studies of disability. In Proctor, J & Smith, D. (eds.) *Geography and Ethics*, London, Routledge, 223–236.

Koskela, H. & Pain, R. 2000. Revisiting fear and place: Women's fear of attack and the built environment. *Geoforum, 31*, 269–280.

Koutsoklenis, A. & Papadopoulos, K. 2011. Olfactory cues used for wayfinding in urban environments by individuals with visual impairments. *Journal of Visual Impairment & Blindness, 105*, 692–702.

Kusenbach, M. 2003. Street phenomenology: The go-along as ethnographic tool. *Ethnography, 4*, 455–485.

Latham, A. 2003. Research, performance, and doing human geography: Some reflections on the diary-photograph, diary-interview method. *Environment and Planning A, 35*, 1993–2017.

Laurier, E. 2010. Cognition and driving. *Mobilities: Practices, paces, subjects*, London, Ashgate.

Laurier, E., Lorimer, H., Brown, B., Jones, O., Juhlin, O., Noble, A., Perry, M., Pica, D., Sormani, P., Strebel, I., Swan, L., Taylor, A. S., Watts, L. & Weilenmann, A. 2008. Driving and 'passengering': Notes on the ordinary organization of car travel. *Mobilities, 3*, 1–23.

Law, J. 2004. *After method: Mess in social science research*, London, Routledge.

Lefebvre, H. 2004. *Rhythmanalysis: Space, time, and everyday life*, London, Continuum.

Lomax, H. 2015. Seen and heard? Ethics and agency in participatory visual research with children, young people and families. *Families, Relationships and Societies, 4*, 493–502.

Lorimer, H. & Lund, K. 2003. Performing facts: Finding a way over Scotland's mountains. *The Sociological Review, 51*, 130–144.

Low, K. E. 2012. The social life of the senses: Charting directions. *Sociology Compass, 63*, 271–282.

Low, K. E. Y. & Kalekin-Fishman, D. 2018. *Senses in cities: Experiences of urban settings*, New York, Routledge.

McDowell, L. 1995. Bodywork: Heterosexual gender performances in city workplaces. In: Bell, D. & Valentine, G. (eds.) *Mapping desire: Geographies of sexualities*, London, Routledge, 75–95.

Macpherson, H. 2009. Articulating blind touch: Thinking through the feet. *The Senses and Society, 4*, 179–193.

Macpherson, H. 2006. Landscape's ocular-centrism–and beyond. In: Tres, B., Tres, G., Fry, G. & Opdam, P. (eds.) *From landscape research to landscape planning: Aspects of integration, education and application*, 12, Dordrecht, Springer, 95–104.

Macpherson, H. 2016. Walking methods in landscape research: Moving bodies, spaces of disclosure and rapport. *Landscape Research, 41*, 425–432.

Manduchi, R. & Kurniawan, S. 2011. Mobility-related accidents experienced by people with visual impairment. *AER Journal: Research and Practice in Visual Impairment and Blindness, 4*, 44–54.

Massey, D. B. 2001. Living in Wythenshawe. In: Borden, I., Kerr, J., Rendell, J. & Pivaro, A. (eds.) *The unknown city: Contesting architecture and social space*, Cambridge, MA, London, MIT, 458–475.

McCormack, D. P. 2005. Diagramming practice and performance. *Environment and Planning D: Society and Space, 23*, 119–147.

_____. 2006. For the love of pipes and cables: A response to Deborah Thien. *Area, 38*, 330–332.

_____. 2010. Remotely sensing affective afterlives: The spectral geographies of material remains. *Annals of the Association of American Geographers, 100,* 640–654.

Merriman, P. 2014. Rethinking mobile methods. *Mobilities, 9,* 167–187.

Middleton, J. 2010. Sense and the city: Exploring the embodied geographies of urban walking. *Social & Cultural Geography, 11,* 575–596.

_____. 2011. "I'm on autopilot, I just follow the route": Exploring the habits, routines, and decision-making practices of everyday urban mobilities. *Environment and Planning A, 43,* 2857–2877.

Middleton, J. & Byles, H. 2019. Interdependent temporalities and the everyday mobilities of visually impaired young people. *Geoforum, 102,* 76–85.

Middleton, J. & Samanani, F. 2021. Accounting for care within human geography. *Transactions of the Institute of British Geographers, 46,* 29–43.

Mitchell, W. J. T. 2002. Showing seeing: A critique of visual culture. In: Mirzoeff, N. (ed.) *The visual culture reader* (2nd ed.), London, Routledge.

Moles, K. 2019. Mobile methods. *SAGE Research Methods Foundations* [available at: http://dx. doi. org/10.4135/9781526421036847435].

Morris, B. & Rose, M. 2019. Pedestrian provocations: Manifesting an accessible future. *Global Performance Studies,* 2.2.

Moser, I. & Law, J. 1999. Good passages, bad passages. *The Sociological Review, 47,* 196–219.

Myers, M. 2011. Walking again lively: Towards and ambulant and conversive methodology of performance and research. *Mobilities, 6,* 183–201.

Nijs, G. & Daems, A. 2012. And what if the tangible were not, and vice versa? On boundary works in everyday mobility experience of people moving into old age: For Daisy (1909–2011). *Space and Culture, 15,* 186–197.

O'Neill, M. & Einashe, I. 2019. Walking borders, risk and belonging. *Journal of Public Pedagogies, 4,* 31–50.

O'Neill, M. & Roberts, B. 2020. *Walking methods: Research on the move,* London, Routledge.

Pain, R. 2000. *Place, social relations and the fear of crime: A review,* Thousand Oaks, CA, Sage.

Pallasmaa, J. 2005. [1991] Lived space: Embodied experience and sensory thought. In: MacKeith, P. (ed.) *Encounters: Architectural essays.* Hameenlinna, Rakennustieto Oy.

Palmgren, A. C. 2018. Standing still: Walking interviews and poetic spatial inquiry. *Area, 50,* 372–383.

Parent, L. 2016. The wheeling interview: Mobile methods and disability. *Mobilities: Mobilities Intersections, 11,* 521–532.

Parr, H. 2007. Collaborative film-making as process, method and text in mental health research. *Cultural Geographies, 14,* 114–138.

Pile, S. 2010. Emotions and affect in recent human geography. *Transactions of the Institute of British Geographers, 35,* 5–20.

Pink, S. 2009. *Doing sensory ethnography,* London, Sage Publications Ltd.

Pink, S., Hubbard, P., O'Neill, M. & Radley, A. 2010. Walking across disciplines: From ethnography to arts practice. *Visual Studies: Walking, Ethnography and Arts Practice, 25,* 1–7.

Porcelli, P., Ungar, M., Liebenberg, L. & Trépanier, N. 2014. (Micro)mobility, disability and resilience: Exploring well-being among youth with physical disabilities. *Disability & Society, 29,* 863–876.

Potter, J. & Wetherell, M. 1987. *Discourse and social psychology: Beyond attitudes and behaviour*, London, Sage.

Pow, C. P. 2017. Sensing visceral urban politics and metabolic exclusion in a Chinese neighbourhood. *Transactions of the Institute of British Geographers, 42*, 260–273.

Pullin, G. 2009. *Design meets disability*, Cambridge, MA, London, MIT.

Ricketts-Hein, J. R., Evans, J. & Jones, P. 2008. Mobile methodologies: Theory, technology and practice. *Geography Compass, 2*, 1266–1285.

Rodaway, P. 1994. *Sensuous geographies: Body, sense and place*, London, Routledge.

Royal Society for Blind People (RSBP). 2014. Youth Manifesto [available at: https://www.yumpu.com/en/document/read/50552052/rlsb-youth-forum-manifesto-28-feb-web].

Sacks, H. 1992. (edited by G. Jefferson) *Lectures on conversation*, Oxford, Blackwell.

Saerberg, S. 2010. "Just go straight ahead": How blind and sighted pedestrians negotiate space. *The Senses and Society, 5*, 364–381.

Saldanha, A. 2002. Music tourism and factions of bodies in Goa. *Tourist Studies, 2*, 43–62.

Sandhu, S. 2007. *Night haunts: A journey through the London night*, London, Artangel/Verso.

Schwanen, T., Hardill, I. & Lucas, S. 2012. Spatialities of ageing: The co-construction and co-evolution of old age and space. *Geoforum, 43*, 1291–1295.

Seamon, D. 1980. Body-subject, time-space routines, and place-ballets. In: Buttimer, A. & Seamon, D. (eds.) *The human experience of space and place*, London, Croom Helm, 148–165.

Shaw, R. 2014. Beyond night-time economy: Affective atmospheres of the urban night. *Geoforum, 51*, 87–95.

Sheller, M. & Urry, J. 2006. The new mobilities paradigm. *Environment and Planning A, 38*, 207–226.

Silver, J., Fields, D., Goulding, R., Rose, I. & Donnachie, S. 2020. Walking the financialized city: Confronting capitalist urbanization through mobile popular education. *Community Development Journal* 56(4), 1–19.

Simmel, G. 1971. *The metropolis and mental life*, Chicago, University of Chicago Press.

Solnit, R. 2014. *Wanderlust: A history of walking*, London, Verso.

Spinney, J. 2009. Cycling the city: Movement, meaning and method. *Geography Compass, 3*, 817–835.

_____. 2011. A chance to catch a breath: Using mobile video ethnography in cycling research. *Mobilities, 6*, 161–182.

_____. 2015. Close encounters? Mobile methods, (post)phenomenology and affect. *Cultural Geographies, 22*, 231–246.

Springgay, S. & Truman, S. E. 2018. *Walking methodologies in a more-than-human world: WalkingLab*, Abingdon, Routledge.

_____. 2019. Walking in/as publics: Editors introduction. *Journal of Public Pedagogies, 4*, 1–12.

Stevenson, N. & Farrell, H. 2018. Taking a hike: Exploring leisure walkers' embodied experiences. *Social & Cultural Geography, 19*, 429–447.

Stone, E. & Priestley, M. 1996. Parasites, pawns and partners: Disability research and the role of non-disabled researchers. *British Journal of Sociology, 47*, 699–716.

Taylor, A., et al. (2010). *Examined life*. [Montréal], National Film Board of Canada [available at: https://www.tarshi.net/inplainspeak/voices-when-sunaura-taylor-and-judith-butler-go-for-a-walk/].

Thien, D. 2005. After or beyond feeling? A consideration of affect and emotion in geography. *Area, 37*, 450–454.

Tolia-Kelly, D. 2006. Affect – An ethnocentric encounter? Exploring the 'universalist' imperative of emotional/affectual geographies. *Area, 38*, 213–217.

Thrift, N. 2004. Intensities of feeling: Towards a spatial politics of affect. *Geografiska Annaler 86 B, 1*, 57–78.

_____. 2007. *Non-representational theory: Space, politics, affect*, London, Routledge.

Urry, J. 2005. The place of emotions within place. In: Davidson, J., Bondi, L. & Smith, M. (eds.) *Emotional geographies*, Aldershot, Ashgate, 77–83.

Vannini, P., Waskul, D. & Gottschalk, S. 2013. *The senses in self, society, and culture: A sociology of the senses*, London/Abingdon, Routledge.

Warren, S. 2017. Pluralising the walking interview: Researching (im)mobilities with Muslim women. *Social & Cultural Geography, 18*, 786–807.

Wetherell, M. 2013. Affect and discourse – What's the problem? From affect as excess to affective/discursive practice. *Subjectivity, 6*, 349–368.

Whittle, R. 2019. Baby on board: The impact of sling use on experiences of family mobility with babies and young children. *Mobilities, 14*, 137–157.

Wilson, E. 1991. *The sphinx in the city*, London, Virago.

Wilson, H. F. 2011. Passing propinquities in the multicultural city: The everyday encounters of bus passengering. *Environment and Planning A, 43*, 634–649.

Wolifson, P. 2016. Encountering the night with mobile methods. *Geographical Review, 106*, 174–200.

Worth, N. 2013. Visual impairment in the city: Young people's social strategies for independent mobility. *Urban Studies, 50*, 574–586.

Wylie, J. 2002. An essay on ascending Glastonbury Tor'. *Geoforum, 33*, 441–455.

Wylie, J. 2005. A single day's walking: Narrating self and landscape on the South West coast path. *Transactions of the Institute of British Geographers, 30*, 234–247.

Zarb, G. 1992. On the road to Damascus: First steps towards changing the relations of disability research production. *Disability, Handicap & Society, 7*, 125–138.

7   Stepping forwards
Some concluding thoughts

About a year into my PhD research I remember finding a flyer in my pigeon hole for a new interdisciplinary journal entitled *Mobilities*. As I read the description of the types of research the journal hoped to publish I began to get excited. The experiential dimensions of urban walking had become central to my research, yet I had often struggled to find an intellectual home for how my thinking was developing. The journal was launched to address a growing body of work across the social sciences, often referred to as the 'mobility turn' (Hannam et al., 2006). This turn represented a shift from understandings of the world that are static and fixed to prioritising the concept of 'mobility' and relational, fluid, and process-orientated approaches. As such, the journal was centred around concerns with mobilities in terms of 'the large-scale movements of people objects, capital and information across the world, as well as the more local processes of daily transportation, movement through public space and the travel of material things within everyday life' (Hannam et al., 2006: 1).

What really grabbed my attention was how the journal was explicitly seeking to engage with, and understand, the complexity associated with people's everyday experiences of mobility. I jumped at the opportunity to attend the official launch of the journal at the AAG Annual Conference in Chicago in 2006. It was there that I began talking with a self-identified 'scholar of mobilities' who happened to ask what I was working on. As I explained that I was seeking to pull together concerns with urban walking across theory and policy into dialogue through my empirical work, they were very quick to proclaim that 'policy is boring'. The implication being: why would anyone want to concern themselves with policy in the context of issues emerging from this growing field of mobilities? This exchange has stayed with me for many years. For me, the troubling nature of such a comment, and similar ones I have encountered since, relates to how academic work translates into spheres beyond academic circles. In this concluding chapter I provide a series of reflections on the arguments I have presented through this book on urban walking in the broader context of this issue.

For a number of years, the above conversation was reflective of what has been framed as a perceived 'transport/mobilities divide' whereby transport studies were considered to be policy relevant yet unimaginative, and mobilities scholarship to be innovative yet overly theoretical (see Cresswell, 2011, 2012, 2014; Goetz et al., 2009; Hall, 2010; Horner and Casas, 2006; Keeling, 2007; Merriman, 2015; Shaw and Hesse, 2010; Shaw and Sidaway, 2011). More recently, there has been a shift away from this dualistic thinking as scholarship has strived even further to emphasise the commonalities, overlaps, and collective understandings of transport and mobility (see Schwanen, 2016). Relatedly, debates framed as to what constitutes 'policy relevant' academic research (see Dorling and Shaw, 2002; Hamnett, 2009; Imrie, 2004) have also diminished in recent years. Such issues now emerge in the context of research impact agendas and how knowledge is valorised, especially in relation to research funding (see Pain et al., 2011; Slater, 2012). Yet, despite these shifts in debate, how different knowledges and experiences translate across different academic fields and policy/practice still warrants further attention.

From 2013 to 2021 I was Course Director of the executive education programme in the Transport Studies Unit at the University of Oxford. The Global Challenges in Transport (GCT) programme is composed of four courses that focus on the most pressing sustainable transport challenges facing planners, researchers, and decision-makers across a range of sectors (see Verlinghieri and Middleton, 2020). At the time of writing, these are smart technologies, climate change, infrastructure, and health and wellbeing. Many of the course participants, who attend from across the globe, are passionately committed to low-carbon mobility futures and the creation of healthy and liveable cities through the promotion of walking and cycling. However, in the eight years I led this programme, I continually observed the disconnect between how walking is understood within urban policy and practice and academic engagements grounded in social and cultural theory. This book has illustrated how walking continues to attract attention across both academic and policy spheres, but how the different bodies of work emerging from each are inherently disconnected and lack a coherent dialogue. Yet, to what extent does/should this matter?

The inclusion and exclusions of everyday urban walking

Throughout the book I have situated my engagements with walking across different disciplinary and policy boundaries and demonstrated the productiveness of working at these interfaces. I have addressed two central concerns: first, how urban space is produced by different walking practices, and second, the forms of inclusions and exclusions that emerge from how a walkable city is understood and practised. I have argued how conceptualising everyday urban walking practices as being socially and materially co-produced needs to be at the centre of addressing such questions. This re-conceptualisation of walking has informed the empirical basis of the book. For example, in Chapter 2, I argued that understandings of urban walking need to move beyond solely focusing on the built environment and traditional wayfinding information and technology. Instead, attention should turn to how both material forms *and* everyday practices co-produce pedestrian infrastructures.

Conceptualising urban walking in this way is also paramount in recognising the significance of the multiple ways in which how people appropriate space on foot matters to our understandings of contemporary urban life. In Chapters 3, 5, and 6, for example, I have examined the differential nature of walking in the city and associated inequalities emerging from the complex inter-relations between class, race/ethnicity, gender, age, and disability. In particular, I have shown how a social politics of walking emerges from the ways in which certain pedestrian types/forms/practices are deemed more important/significant than others. I have also argued for walking bodies to be at the centre of understandings of urban walking. In Chapter 6, I propose that engaging with the embodied, material, and technological relationality of walking disrupts self-evident notions of walkability. In doing so, I contend that greater account needs to be given to differentiated mobile bodies. I draw upon critical reflections on the use of walking methods in relation to notions of accessibility and the 'demands of the method' (Warren, 2017: 802) to emphasise the significance of differentiated mobile bodies to understandings of urban pedestrian practices.

The lived experiences of the walking body

The empirical content of this book has been the culmination of several years' work on everyday walking that has been born out of projects based in the UK. As researchers, our engagement with different projects is often serendipitous, emerging out of chance encounters, our domestic circumstances/constraints and against the backdrop of a neoliberalising Higher Education sector where funding opportunities are simultaneously becoming increasingly competitive and scarce. I make these points as a means of providing some context to the Eurocentric nature of the empirical contents of the book. Yet, despite this empirical focus, the arguments I have made concerning the significance of engaging with the socially differentiated nature of everyday walking in relation to contemporary urban life resonate both within, and beyond, Europe and North America.

On 25 May 2020, the brutal murder of George Floyd in Minneapolis, Minnesota, by a US police officer, brought renewed attention to the racial inequalities that continue to be experienced by black people in both the United States and beyond. The incident sparked global protests led by the Black Lives Matter movement, demanding an end to years of oppression and systemic racism. A friend once told me that he never wears a hoodie out in public, never walks closely behind someone, and always avoids walking past, or standing too close to, a person withdrawing money out of a cashpoint. His actions stem from his fear of being criminalised simply due to him being a black male (see Rabinowitz, 2015, on the criminalisation of being black) and illustrate the systemic and structural racism embedded in how everyday mobilities are lived, practised, and experienced. The murder of George Floyd, and countless others, and my friend's experiences, which are tragically far from unique, are stark reminders of the racial inequalities that remain deeply embedded within our society. In the context of walking, there is a pressing need to unlearn the embodiments of walking and what we take walking to be. For example, what are the embodied experiences, and associated implications, of 'walking while black' (Nicholson, 2016) or walking as disabled, as a woman, as a child, as an older person, or in poverty? And how do such experiences intersect?

In recent years there have been calls to decolonise our understandings of mobility (Kwan and Schwanen, 2016) and more specifically walking practices (Murali, 2017). For example, in the context of Shenzhen, China, Chan et al. (2020) question the transferability of 'Western' understandings of walking to 'developing cities'. They illustrate how bodily fitness was of limited concern to their research participants. Instead, walking was considered in the context of the preventative principles of Chinese medicine and its potential for maintaining harmony between mind, body, and environment. Elsewhere, alongside Ersilia Verglinghieri, I have argued for the importance of contesting the universalising dominance of 'Western' paradigms, methods, themes, and solutions to transport challenges (see Verlinghieri and Middleton, 2020). The need to decolonise walking knowledges is an important part of this call, particularly in relation to the walking body. This book can be read as the beginnings of such a move through its centring of walking bodies and the socially differentiated experiences of everyday walking. Close attention to the lived experiences of everyday pedestrian practices allows a re-thinking of what walking actually is.

The growing body of work in critical walking studies acknowledges the pressing need to 'complicate and rupture the White-cis-hetero-ableist-patriarchal canon of walking scholarship' (Springgay and Truman, 2019: 2). A range of research projects and powerful performative artistic pieces have developed within this field (see for example Arora, 2020; Johnson, 2019; Somerville et al., 2019) that contrasts with many engagements with walking that emerge from a position of class, raced, and/or gendered privilege. Yet such work remains firmly rooted in the arts and humanities while related social science engagements are dominated by concerns with walking as a methodology (see Bates and Rhys-Taylor, 2017). This leaves a key question of how the importance of decolonising walking practices and centring the socially differentiated nature of the lived experiences of pedestrian practices translates beyond these arenas. For example, how might the policy makers and practitioners who attend the GCT programme engage with such concerns?

Appropriating space on foot and the 'right to mobility'

This book has engaged with concerns with mobility justice through considering the 'right to mobility' in terms of people's differential rights and experiences to move in, through, across, and between different places and how these are fundamental to understandings of everyday urban walking practices. In particular, I have drawn attention to how 'who' can walk in certain urban spaces, and 'where', is mediated by a series of power relations that are socially and materially co-produced. This has made visible the significance of the right to mobility in the context of everyday pedestrian practices in relation to 'who' and for 'whom', walkable cities are imagined and practised while challenging the status quo of walking beyond the white, male, heterosexual, non-disabled body. The concept of mobility justice will be familiar to most transport practitioners. However, as Verlinghieri and Schwanen (2020) highlight, any mobility justice agenda needs to move beyond a 'universal disembodied subject' and take account of 'the differential voices, knowledges, experiences, abilities and rhythms of the actors that inhabit particular spaces and places'.

The assumptions that walking is an inherently 'just' mobile practice for everyone need to challenged. For example, in the opening of Chapter 1, I describe the 2.5-mile journey on foot of Courtney and her mum and younger brother to the local foodbank. They did not want to be walking as they were laden with heavy bags of food but they could not afford the bus fare. It is difficult to frame their journey on foot in line with dominant framings in both academia, policy, and practice in relation to the emancipatory and democratising potential of walking. It is therefore imperative that challenges to such assumptions are central to the ways in which everyday urban walking is engaged with and understood across theory, policy, and practice. Equally, the importance and opportunities of situating future work at the interfaces of these fields, as opposed to operating in silos, should not be underestimated. This book is an envisioning of such an interface.

Through this book I have invited the reader to think differently about everyday urban walking. In particular, I have sought to challenge the taken-for-grantedness of walking and the assumptions which are deeply embedded in how walking is imagined, positioned, and promoted in contemporary urban life. It is through this approach that we become more attuned to the inclusions and exclusions that emerge from our everyday pedestrian practices. In particular, I argue that a far greater reflection is required to the routine, habitual, and everyday experiences of those people who actually walk in city spaces; and how the nuanced politics of walking practices are not only fundamental in recognising, and encouraging, more progressive forms of urban encounters but also the differential nature of walking practices and associated implications. Concerns with 'how' people walk in terms of their lived experiences, as opposed to prioritising concerns with 'why', can assist in providing more in-depth understandings of the complexity of walking practices, and how we conceive of walkability, than is currently engaged with in the urban, health, and transport policy arena of research and practice. There is a need to engage more fully with what it is 'to do' walking. In other words, taking the experiential dimensions of walking practices to be as significant as the built environment in engaging with the complexity of the relationship between walking and urban space and how the walkable city, through the 'simple act of walking', is understood and conceptualised.

References

Arora, S. 2020. Walk in India and South Africa: Notes towards a decolonial and transnational feminist politics. *South African Theatre Journal*, 1–20.

Bates, C. & Rhys-Taylor, A. 2017. *Walking through social research*, London, Routledge.

Chan, E. T. H., Schwanen, T. & Banister, D. 2020. People and their walking environments: An exploratory study of meanings, place and times. *International Journal of Sustainable Transportation*, 1–12.

Cresswell, T. 2011. Mobilities I: Catching up. *Progress in Human Geography*, 35, 550–558.

_____. 2012. Mobilities II: Still. *Progress in Human Geography*, 36, 645–653.

_____. 2014. Mobilities III: Moving on. *Progress in Human Geography*, 38, 712–721.

Dorling, D. & Shaw, M. 2002. Geographies of the agenda: Public policy, the discipline and its (re)'turns'. *Progress in Human Geography*, 26, 629–641.

Goetz, A. R., Vowles, T. M. & Tierney, S. 2009. Bridging the qualitative-quantitative divide in transport geography. *The Professional Geographer*, 61, 323–335.

Hall, D. 2010. Transport geography and new European realities: A critique. *Journal of Transport Geography*, 18, 1–13.

Hamnett, C. 2009. The new Mikado? Tom Slater, gentrification and displacement. *City*, 13, 476–482.

Hannam, K., Sheller, M. & Urry, J. 2006. Editorial: Mobilities, immobilities and moorings. *Mobilities*, 1, 1–22.

Horner, M. W. & Casas, I. 2006. An introduction to assessments of research needs in transport geography. *Journal of Transport Geography*, 14, 228–229.

Imrie, R. 2004. Urban geography, relevance, and resistance to the "policy turn". *Urban Geography*, 25, 697–708.

Johnson, W. 2019. Walking Brooklyn's redline: A journey through the geography of race. *Journal of Public Pedagogies*, 4, 209–216.

Keeling, D. J. 2007. Transportation geography: New directions on well-worn trails. *Progress in Human Geography*, 31, 217–225.

Kwan, M.-P. & Schwanen, T. 2016. Geographies of mobility. *Annals of the American Association of Geographers*, 106, 243–256.

Merriman, P. 2015. Mobilities, crises, and turns: Some comments on dissensus, comparative studies, and spatial histories. *Mobility in History*, 6, 20–34.

Murali, S. 2017. A manifesto to decolonise walking: Approximate steps. *Performance Research: On Proximity*, 22, 85–88.

Nicholson, J. A. 2016. Don't shoot! Black mobilities in American gunscapes. *Mobilities: Mobilities Intersections*, 11, 553–563.

Pain, R., Kesby, M. & Askins, K. 2011. Geographies of impact: Power, participation and potential. *Area*, 43, 183–188.

Rabinowitz, P. 2015. Street crime from Rodney King's beating to Michael Brown's shooting. *Cultural Critique*, 143–147.

Schwanen, T. 2016. Geographies of transport I: Reinventing a field? *Progress in Human Geography*, *40*, 126–137.

Shaw, J. & Hesse, M. 2010. Transport, geography and the 'new' mobilities. *Transactions of the Institute of British Geographers*, *35*, 305–312.

Shaw, J. & Sidaway, J. D. 2011. Making links: On (re)engaging with transport and transport geography. *Progress in Human Geography*, *35*, 502–520.

Slater, T. 2012. Impacted geographers: A response to pain, Kesby and Askins. *Area*, *44*, 117–119.

Somerville, M., Tobin, L. & Tobin, J. 2019. Walking contemporary indigenous songlines as public pedagogies of country. *Journal of Public Pedagogies*, *4*, 13–27.

Springgay, S. & Truman, S. E. 2019. 2019. Walking in/as publics: Editors introduction. *Journal of Public Pedagogies*, *4*, 1–12.

Verlinghieri, E. & Middleton, J. 2020. Decolonising and provincialising knowledge within the neoliberal university? The challenge of teaching about sustainable transport. *Journal of Transport Geography*, 88.

Verlinghieri, E. & Schwanen, T. 2020. Transport and mobility justice: Evolving discussions. *Journal of Transport Geography*, 87.

Warren, S. 2017. Pluralising the walking interview: Researching (im)mobilities with Muslim women. *Social & Cultural Geography*, *18*, 786–807.

Epilogue

In December 2019, a growing number of people began to be hospitalised with severe respiratory problems and pneumonia in the city of Wuhan in the Hubei province of China. The earliest cases of Coronavirus (COVID-19) have been traced to seafood and animal markets in Wuhan, yet its rapid spread across the globe has been by person-to-person contact mainly via air droplets expelled through the cough of those infected (although many people have carried and transmitted the virus but remained asymptomatic). At the time of writing, the virus has claimed over 3 million lives globally. The COVID-19 pandemic has had profound effects on our everyday mobilities at a global and local scale in ways many could never have anticipated. As the virus spread, most countries were eventually 'locked down' with stringent social distancing measures in attempts to minimise the spread of the virus by telling people to stay at home. As these enforcements to restrict everyday mobility beyond the home were put in place, questions such as 'is it safe to go for a walk?' circulated on social media and became central topics of numerous newspaper and magazine articles. As such, the very essence of what walking is and what it is to go for a walk became a prominent concern. Some used this as an opportunity to continue promoting the significance of 'safe walking and cycling environments' for healthy living. For example, a group of UK transport researchers called on the government in an open letter to 'enable safe walking and cycling during the COVID-19 pandemic' (Aldred, 2020). Others emphasised the importance of networked, pedestrianised streets and communities (see James, 2020). Several countries, including Belgium and Germany, temporarily closed roads to allow people to walk 2 m apart, while some governments, such as the UK and New Zealand, announced funding packages for walking and cycling infrastructures.

Like many other countries, the UK[1] eventually went into lockdown. On 23 March 2020, UK government advice was now to only go outside for food, health reasons, or work if you could not work from home. When out to stay 2 m away from other people and to not meet others including friends and family (GOV.UK, 2020). People were allowed to go out for exercise once a day but this was to take place in open spaces near where they lived. This specific advice on the frequency and location of daily exercise only subsequently emerged following the controversies surrounding swathes of people visiting parks and beaches despite social distancing measures. On 23 June 2020, the UK government announced the gradual easing of lockdown measures in England, including reducing social distancing recommendations to a minimum of 1 m (3 feet) where 2 m (6 feet) is not possible. This easing of restrictions was criticised by many as a political decision driven by concerns relating to the recovery of the economy at the expense of people's health. At the time of writing, many countries are in the midst of contending with a second or third wave of the virus. The approach to imposed restrictions, in places such as the UK and France, had taken the form of localised actions responding to geographical differences in infection levels. However, infection rates continued to rise and in several countries, including the UK, a second and then third national lockdown was imposed. In the context of walking, despite the constantly changing parameters of restrictions and the hope that a new vaccine provides, I am convinced that the very nature of pedestrian practices will remain altered for the foreseeable future.

My family are fortunate to have access to urban green space very close to our house. However, although the number of people being out significantly decreased during the first lockdown, it was still necessary to navigate a range of different encounters in attempts to follow the guidelines on appropriate social distancing of being 2 m apart. After a few days, I began to find these daily exercise excursions, which were meant to contribute to our mental and physical well-being, quite stressful. I found myself continually barking instructions to 'stay to the left' and 'don't go too close!' to an already confused four- and seven-year-old. The phrase 'herding cats' continually sprang to mind. I felt a continual pressure and weight of societal responsibility on such walks that began to make them far from enjoyable. However, throughout all of these walks and associated encounters, one in particular has stayed with me.

Towards the end of the first week of lockdown in the UK, I took my two young children out for this daily permitted exercise. As we walked towards the gate to the nature reserve, a reclaimed landfill site, I saw two cyclists approaching from the other side. An impromptu race began as my children began running at speed to the gate, followed by my repeated shouts for them to 'stop' and 'wait', growing with intensity at each cry. They eventually stopped at the gate as the masked and gloved cyclists on the other side speedily retreated away with looks of what I can only describe as pure fear. I attempted to engage politely with the woman and her teenage son by apologetically explaining that my youngest was only four and did not fully understand the current situation. She ignored me and continued to retreat while pleading with her son to follow her. I continued my attempt to explain, but, again, this was met with no acknowledgement but a further fearful retreat. It was at this point that I felt a surge of frustration and anger rise up with an intensity that shocked me. I then found myself angrily asking her what I was supposed to do: 'he's only four and doesn't understand. If you are so concerned and scared then stay at home. You are out, we are also allowed to be out, please be more understanding'.

This book is not an appropriate space to discuss the 'rights and wrongs' of this encounter in relation to my frustrated outburst. Upon subsequent reflection, I realise that both of us were probably exhibiting stress and fear in relation to the situation we found ourselves in. I also have no doubt that numerous similar exchanges occurred in many different countries during these periods of lockdown. Conflicts between pedestrians and cyclists are not unusual, particularly in urban space (see Middleton, 2018). However, in the wider context of the COVID-19 pandemic, the tension that unfolded through such an encounter was of particular significance in further exemplifying the central arguments of this book in relation to taking more seriously the socially differentiated nature of everyday walking practices. In this societal repositioning of walking during the COVID-19 outbreak, pedestrian bodies and our differential experiences of walking became centre stage. An everyday politics emerges from the restrictions imposed on us by the virus as to who is able to walk or not (older people, those with vulnerable health, etc.), and associated expectations of what it is to be a responsible walker maintaining acceptable social distancing. While critiques were made of how this daily allowance of exercise was overwhelmingly focused on adults without acknowledging the different needs of children and how they use outdoor space (see Russell and Stenning, 2020). The rhythms of walking become centralised to our understandings of everyday mobility practices, including when we walk, how long for, and, most significantly, the rhythms of walking bodies as they navigate to avoid unnecessary encounters.

Certainly, in the UK, many photos began to appear on social media of tired dogs who were being taken out for yet another walk as a means of justifying the recommended amount of daily exercise. Dog ownership has also increased with 'pandemic puppies' becoming new additions to many households. Yet, these experiences of walking during the COVID-19 pandemic in the UK sit in stark contrast to the traumatic events that unfolded as the Indian government announced a 21-day lock down on 21 March 2020. As migrant workers fled the cities back to their villages, there was no space on the few overcrowded buses still operating. This left thousands of workers, who now had no work, little choice but to walk, often hundreds of miles home. Many died on these treacherous journeys on foot, with others arriving at villages to find they were not welcome due to fears of them bringing the virus with them. The impact of the virus thus highlights even more acutely how the benefits and positive experiences frequently associated with walking are far from universal experiences and that walking is not one thing to all people.

As the COVID-19 crisis intensified, reflections were made about the mobilities associated with such a global pandemic. For example, Cresswell (2020) published a blog post on the Mobile Lives Forum where he drew upon his previous work on turbulence to emphasise how COVID-19 revealed much that is wrong with the ways we move. In particular, Cresswell highlights the differentiated nature of mobility in relation to the virus. While the thoughts of Sheller (2020) turned to the aftermath of what she terms the 'mobility shock' of COVID-19 and how 'changing the ways we do mobilities' to more socially equitable ways are essential to the recovery of from the virus. As both Cresswell and Sheller highlight, through this devastating global health pandemic our differentiated experiences of everyday mobility have become even more visible. Yet what does this mean for walking? Yes, there are the well-rehearsed concerns in relation to low-carbon transitions, more people being encouraged to exercise, and the promotion of liveable, healthy streets and communities. For example, Lyons (2020) has reflected upon the opportunities the pandemic has opened up for promoting walking while contending that 'it's simply a behaviour (with the right encouragement and improvements to the walking environment) that needs to be performed more' (3). Yet it is the very taken-for-grantedness of what it is to 'simply' go for a walk that needs to be called into greater question. It is, in fact, anything but simple and our differentiated pedestrian experiences are of vital importance for understanding the complexity of everyday walking practices. In particular, our embodied experiences, associated walking 'skills'/capacities, and the inclusions and exclusions that emerge from everyday pedestrian practices matter greatly to how everyday walking is positioned and promoted in urban imaginaries. For a 'walkable city' is one that recognises the diversity of our pedestrian experiences and their significance for understanding the complexity of contemporary urban life.

Note

1. The UK government is only responsible for restrictions in England. Health is a devolved responsibility with Scotland, Wales, and Northern Ireland responsible for their own public health policies.

References

Aldred, R. 2020. Open letter from public health and transport researchers calls on government to support safe walking and cycling (in terms of infection & injury risks) during the COVID-19 pandemic, *Twitter*, 17 March [available at: https://twitter.com/RachelAldred/status/1239987645050245121].

Cresswell, T. 2020. Mobility: The lifeblood of modernity and the virus that threatens to undo it, Mobile Lives Forum [available at: https://en.forumviesmobiles.org/2020/03/18/mobility-lifeblood-modernity-and-virus-threatens-undo-it-13266].

GOV.UK (2020) Staying at home and away from others: social distancing. Updated 1 May 2020 [available at: https://www.gov.uk/government/publications/full-guidance-on-staying-at-home-and-away-from-others/full-guidance-on-staying-at-home-and-away-from-others].

James, S. 2020. What streets should Vancouver close for walking, rolling & physical distancing? *Price Tags* [available at: http://pricetags.ca/2020/03/30/what-streets-should-vancouver-close-for-walking-rolling-physical-distancing/].

Lyons, G. 2020. And don't forget walking: taking steps out of the pandemic. Write-up of the fifth PTRC Fireside Chat examining the transport implications of the pandemic, September [available at: https://www.linkedin.com/pulse/dont forget-walking-taking-steps-out-pandemic-glenn-lyons/and also at: https://drive.google.com/file/d/1IesaqlnAO4IVDWQs-wNcKTmFIgWX66FT/view?usp=sharing].

Middleton, J. 2018. The socialities of everyday urban walking and the 'right to the city'. *Urban Studies*, *55*, 296–315.

Russell, W. & Stenning, A. 2020. Beyond active travel: Children, play and community on streets during and after the coronavirus lockdown. *Cities & Health*, 1–4.

Sheller, M. 2020. Some thoughts on what comes after a mobility shock. *Critical Automobility Studies Lab blog* [available at: https://cas.ihs.ac.at/some-thoughts-on-what-comes-after-a-mobility-shock/].

Index

Printed in the United States
by Baker & Taylor Publisher Services

Printed in the United States
by Baker & Taylor Publisher Services